T0334594

NON-MONOTONIC APPROACH TO ROBUST H_∞ CONTROL OF MULTI-MODEL SYSTEMS

NON-MONOTONIC APPROACH TO ROBUST H_∞ CONTROL OF MULTI-MODEL SYSTEMS

Jiwei Wen

Alireza Nasiri

Sing Kiong Nguang

Dhafer J. Almakhles

ACADEMIC PRESS
An imprint of Elsevier

Academic Press is an imprint of Elsevier
125 London Wall, London EC2Y 5AS, United Kingdom
525 B Street, Suite 1650, San Diego, CA 92101, United States
50 Hampshire Street, 5th Floor, Cambridge, MA 02139, United States
The Boulevard, Langford Lane, Kidlington, Oxford OX5 1GB, United Kingdom

Notices

Knowledge and best practice in this field are constantly changing. As new research and experience broaden our understanding, changes in research methods, professional practices, or medical treatment may become necessary.

Practitioners and researchers must always rely on their own experience and knowledge in evaluating and using any information, methods, compounds, or experiments described herein. In using such information or methods they should be mindful of their own safety and the safety of others, including parties for whom they have a professional responsibility.

To the fullest extent of the law, neither the Publisher nor the authors, contributors, or editors, assume any liability for any injury and/or damage to persons or property as a matter of products liability, negligence or otherwise, or from any use or operation of any methods, products, instructions, or ideas contained in the material herein.

Library of Congress Cataloging-in-Publication Data
A catalog record for this book is available from the Library of Congress

British Library Cataloguing-in-Publication Data
A catalogue record for this book is available from the British Library

ISBN: 978-0-12-814868-6

For information on all Academic Press publications
visit our website at https://www.elsevier.com/books-and-journals

Working together
to grow libraries in
developing countries

www.elsevier.com • www.bookaid.org

Publisher: Mara Conner
Acquisition Editor: Sonnini R. Yura
Editorial Project Manager: Gabriela D. Capille
Production Project Manager: R.Vijay Bharath
Designer: TBD

Typeset by VTeX

Contents

5. Stability and H_∞ Control of Discrete-Time Switched Systems via One-Step Ahead Lyapunov Function Approach

6. Stability, l_2-Gain and Robust H_∞ Control of Switched Systems via Multistep Ahead Nonmonotonic Approach

7. Robust H_∞ Filtering for Average Dwell-Time Switched Systems via a Nonmonotonic Function Approach

8. Dissipative Dynamic Output Feedback Control for Switched Systems via Multistep Lyapunov Function Approach

9. Robust H_∞ Control of Discrete-Time Nonhomogenous Markovian Jump Systems via Multistep Ahead Lyapunov Function Approach

10. Robust H_∞ Filtering of Nonhomogeneous Markovian Jump Delay Systems via N-Step Ahead Lyapunov–Krasovskii Function Approach

11. Conclusions and Future Work

Preface

For a long time, the control theory of multimodel systems such as T-S fuzzy systems, nonhomogenous Markovian jump systems, switched systems, etc., is established based on the monotonicity requirement of Lyapunov functionals (LFs) over an infinite-time horizon. Inevitably, the monotonicity requirement of LFs introduces conservatism in the design. This book focuses on the robust analysis and synthesis problems for multimodel systems based on nonmonotonic LF approaches. By fully considering the diversity of the switching laws, the multistep time difference, the multistep prediction, and the expansion of system dimension, the nonmonotonic LF can be properly constructed. The LF's evolution is allowed to increase locally but with an average decrease in every few steps. This nonmonotonic LF approach enlarges the stability region and improves the control performance. The focus of the book has been placed on H_∞ state feedback control, H_∞ filtering, and H_∞ output feedback control for multimodel systems via nonmonotonic LF approaches.

This book considers three classes of complex systems. The first class is the T-S fuzzy model, which is modeled by a set of fuzzy rules with linearized subsystems. The membership functions are used to connect the linear submodels, and the so-called T-S fuzzy model can be employed to approximate any nonlinear system. The second class is the switched system, which is described by a group of indexed subsystems represented by differential or difference equations with a switching rule governing them. Such a switching rule could be time-dependent or state-dependent. Switched systems usually are studied from temporal case or spatial case. The last class is the Markovian jump system, which can effectively model dynamic systems involving stochastic switching subject to a Markov chain.

Several basic issues will be addressed in this book, including stability analysis, dissipativity analysis, controller design, filter design, etc. Firstly, we devote our attention on T-S fuzzy models with modeling uncertainties. By properly constructing LF with nonmonotonic behavior, the criteria for stability analysis, stabilization, and filtering are presented and then are extended to the case of dynamic output feedback control of nonlinear systems. Secondly, we focus on arbitrary switched systems and average dwell-time switched systems. Methodologies that can effectively handle control and filtering problems with less conservatism are developed by allowing the LF to increase both at the switching instant and during the running time of each subsystems. Finally, we deal with Markovian jump systems governed by time-varying transition probabilities. The concept of

N-step ahead approach is developed such that the stability problem can be solved via a finite number of conditions. The systems with time-delay, which are handled by Lyapunov–Krasovskii function approach, are also investigated.

This book aims at providing an overview of the recent research advances on multimodel systems via nonmonotonic LF approach. It can be used in undergraduate and graduate study and is also suitable as a reference for engineers and researchers in system and control field. Prerequisite to reading this book is elementary knowledge on mathematics, matrix theory, probability, convex optimization techniques, and control system theory.

Wuxi, China
Auckland, New Zealand
December 2018

Jiwei Wen
Alireza Nasiri
Sing Kiong Nguang
Dhafer J. Almakhles

Acknowledgments

We would like to express our deep appreciation to those who have been directly involved in various aspects of the research leading to this book. Special thanks go to professor Peng Shi from University of Adelaide in Australia, Professor Xudong Zhao from Dalian University of Technology in China, Professor Akshya Swain from the University of Auckland, Professors Fei Liu, Xiaoli Luan, and Jun Chen from Jiangnan University in China for their valuable suggestions, constructive comments, and support. We also extend our gratitude to many colleagues who have offered technical support and encouragement throughout this research. In particular, we would like to acknowledge the contributions from Yun Xie, Ruichao Li, Qian Feng, Howard Wang, and Aaron. Finally, we are especially grateful to our families for their never-ending encouragement and support when it deemed necessary.

This book was supported in part by National Natural Science Foundation of China (nos. 61203126, 61573069, 61773131, 61773183, 61722306, U1509217), the 111 Project (no. B12018), and the Australian Research Council (DP170102644).

1

Introduction

1.1 NONMONOTONIC APPROACH AND MULTIMODEL SYSTEM

1.1.1 Nonmonotonic Approach

Modern control theory involves many research fields with a set of rigorous analysis and synthesis methods. In control systems theory, stability analysis is the foundation of almost all approaches. By taking the system uncertainties into consideration, the so-called robust stability is also a hot research topic in the last twenty years. The concepts of the asymptotic stability [1], global uniform asymptotic stability (GUAS) [2], stochastic stability [3], etc. usually aim at the equilibrium points of the dynamical systems. The depth and breadth of their theoretical developments are far beyond solving specific problems of certain control systems. As a sufficient condition, Lyapunov stability is a simple and straightforward approach to address the stability analysis problem by properly choosing or constructing a Lyapunov function (LF). However, from an engineering point of view, it inevitably introduces conservativeness to some extent. Therefore, many research efforts are devoted to conservatism reduction problem, for example, designing parameter-dependent LF [4], discussing the necessity condition [5], etc. Some of these approaches successfully reduced the conservatism to a certain extent; other approaches got the analysis results but brought severe difficulties to the controller synthesis. However, none of these approaches fundamentally solved all the analysis and synthesis problems of the controlled systems. Generally, it is believed that if the framework of Lyapunov stability based on the equilibrium point is not breached, then the excessive search for the necessary conditions will gradually remove the research from the engineering background. As a matter of fact, many engineering problems can be solved only by sufficient conditions.

The nonmonotonic Lyapunov function (NLF) is a relaxed version of LF and is based on such a simple idea: consider a proper LF $V > 0$ and $V \to 0$

eventually; a natural question arises: Is the *monotonic* decreasing of V a necessary condition to guarantee $V \to 0$ as $t \to 0$?. Actually, some pioneering work already proposed NLFs in the continuous-time domain when Butz [6] tried to replace $\dot{V} < 0$ by $\ddot{V} < 0$ to obtain a sufficient condition of GUAS. The obtained condition in [6], however, cannot be transformed into a convex optimization problem. Subsequently, Yorke [7] gave convex conditions to address the Lyapunov stability, which cannot guarantee the GUAS. Based on the fundamental work of [6,7], Aeyels and Peuteman [8] developed an efficient approach, i.e., allowing the LF to occasionally increase on several small intervals. However, such an approach cannot achieve controller synthesis, and also the paper has not given detailed solutions in the discrete-time domain.

Since 2008, the NLF approach, which is aiming at the discrete-time nonlinear system, starts to penetrate into the research front line [9]. The main idea is to obtain a GUAS criteria with less conservatism by allowing the LF to increase locally within several sampling period. For the T-S fuzzy model, Derakhshan et al. employed NLF using only 2-samples variations $V(k + 2) - V(k) < 0$. This approach allowed LF to increase locally at $k + 1$, which enlarges the stability region for a class of nonlinear systems. The state feedback synthesis [10], observer-based synthesis [11], and robust H_2 synthesis [12] have also been achieved. The authors in [13–16] extended the results to the general case $V(k + N) - V(k) < 0$, which allowed LF to increase in several sampling periods between the sampling point k and $k + N$, to further reduce the conservativeness when obtaining the stable region of nonlinear systems. The corresponding nonquadratic state feedback stabilization [13], robust H_∞ state feedback control [14], guaranteed cost control [15], and robust dynamic output feedback control [16] have been intensively studied using the general case of NLF with N-sample variations. The studies of NLF approach for T-S fuzzy model is also an important part of this book.

1.1.2 Multimodel System

The real complex system usually contains strong nonlinearities and encounters bad scenarios such as parameter mutations and working condition changes. The synthesis for the multimodel systems (MMSs) provides an effective method for the control of complex systems. The multimodel control system is typically composed of three main parts: a multimodel set, a controller set, and a control synthesis algorithm (also referred to as a switching algorithm). The system structure of MMS is shown in Fig. 1.1.

The basic principle of these three parts in multimodel control system can be summarized as follows. A multimodel set is constructed, as a combination of simple models, to approximate the original complex system. The combination of these simple models is designed by a simple fixed mode of a nonlinear system and thereby is used to represent different

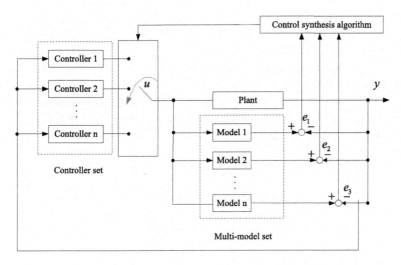

FIGURE 1.1 Multimodel control system.

working conditions, operation processes, etc. Next, a corresponding controller set is designed, as a combination of subcontrollers, to meet the requirements specification of the control system. The difference between the real system and the multimodel set is then processed by the control synthesis algorithm *either* to let the controller set switches between the subcontrollers when there is only a single subcontroller in control action *or* to compute the weighting coefficients for subcontrollers when there are more than one subcontroller in control action. As a result, a global control for the complex nonlinear systems can be achieved using MMSs. Examples of MMSs are T-S fuzzy systems, switched systems, nonhomogenous Markovian jump systems, and artificial neural network.

The main objective of this book is to apply NLF to reduce the conservatism in different types of MMSs, namely T-S fuzzy systems, switched systems with average dwell-time switching, and Markovian jump systems. Throughout the book, robust filtering and control strategies are developed using the NLF approach, which certainly can be used in tackling many control problems in different complex industrial systems.

1.1.2.1 T-S Fuzzy Model

Over the past twenty years, Takagi–Sugeno (T-S) fuzzy models have been successfully utilized to approximate nonlinear systems [17–20]. A T-S fuzzy model of the nonlinear system is constructed as a combination of linear models with nonlinear scalar membership functions. This feature has inspired many researchers to use this model for control, stabilization, and filtering of nonlinear systems represented by T-S models. Linear matrix inequalities (LMIs) have been used extensively to design state feedback control [21–23], static output feedback control [24], dynamic output feedback control [23,25], robust H_∞ control [26–29], H_∞ filtering [30–33], reliable control [34–36], and guaranteed cost control [37–40]. The main drawback associated with the aforementioned methods is that the selected LF must work for all linear subsystems of the T-S fuzzy model. Furthermore, since the membership functions are usually excluded from final stability conditions, this gives rise to sufficient (conservative) conditions in the form of LMIs. These LMIs are often too conservative, and sometimes it may not be possible to ensure stability for highly nonlinear complex systems with a large number of rules. Thus, finding an LF is a challenging area of research.

In recent years, various types of LFs have been proposed for T-S fuzzy models. The first and easiest choice in deriving stability and stabilization is the common quadratic LF [21–23]. However, in many cases, this has been found to be very conservative. Some authors [41–46] have judged stability from a T-S fuzzy model of nonlinear systems by using piecewise quadratic LFs of different forms, which is a much richer class of LFs and may lead to less conservatism than common quadratic LFs. Another type of LFs, which has been significantly used in the past, is fuzzy LFs [47–51]. In a fuzzy LF, the membership functions are often employed to facilitate the stability analysis and reduce the conservativeness. A polynomial LF [52–54] has also been employed to design controllers for T-S fuzzy models giving less conservative results compared to a quadratic LF. Despite all these advances, conservatism is still a problem, and more studies are required.

In real control systems, system performance and operation may degrade because of unknown inputs, such as disturbances or system uncertainties. A robust state feedback controller has been designed using N-sample variations [13]. However, the disturbance effect has not been investigated and still remains a challenge in using NLFs. This motivated the current research seeking to minimize the effect of disturbance by designing an H_∞ controller using an NLF. In addition, a new form of NLF is associated in the design, which reduces the conservatism in comparison with [13] in the case of unperturbed systems. Furthermore, in many physical systems, it may be difficult and costly to measure all of the system states, and only the outputs of the system are available. Therefore, a design for robust H_∞ filtering and output feedback control using NLF is developed.

1.1.2.2 *Switched System With Average Dwell-Time Switching*

The existing results based on the NLF are almost limited in T-S fuzzy models. The reason for this is that the stability criteria developed for T-S fuzzy model must be satisfied for all arbitrary nonlinear membership functions, which can be expected to reduce their conservatism. However, unlike T-S fuzzy model, we found that the LF with a switched system suddenly increase at the switching instants as can be observed in dwell-time (DT) switching [55–57], average dwell-time (ADT) switching [58–62], and Markovian switching [63–65]. Asynchronous switching [66–68] between the plant object and controller/filter also leads to a nonmonotonic evolution process. As a result, the energy function of the switching systems tends to have a nonmonotonic nature in the evolution process. Therefore, if we directly extend the existing NLF method of T-S fuzzy model to switched systems, then some special difficulties can be encountered. For example, the evolution process of the nonmonotonic functional characterization is not clear. The exponential attenuation law is difficult to be obtained. It is a challenging task to give quantitative analysis, design results, and so on. Therefore, it is necessary to carry out further research on the switching systems, especially on some systems with special switching mechanism.

Before deeply understanding what switched systems are, it is usually suggested to know about hybrid systems, which cover switched systems as a particular case. A hybrid system is a system that consists of coupled continuous-time dynamics and discrete-time events interacting with each other. Concerned by both control and computer communities, hybrid systems are usually studied in two different ways. Computer scientists commonly focus on the discrete-time events of hybrid systems and simplify the continuous-time dynamics for the sake of designing computer programs, which based on the discrete logics. On the contrary, the researchers in the area of systems and control engineering take care about the continuous-time dynamics of hybrid systems and abstract evolutions of discrete-time events to a set of supervisory switching laws. Therefore, from the control point of view, such systems are referred to as switched systems. They are usually described by a group of indexed subsystems represented by differential or difference equations with a switching law governing among them. For formal definitions of hybrid systems and their common applications, readers can refer to [69] and the references therein.

From a different point of view, the switched system has many kinds of classification. For example, if we classify the switched system according to the switching laws, then they can be classified into two categories.

1. State-dependent switching and time-dependent switching: In the state-dependent switched systems, the space has been partitioned into subspaces with finite or infinite numbers when each subsystem is operating in the corresponding subspace. When the system state achieves the switching surface, the whole system switches to another subsystem operating in another corresponding subspace. Therefore, the state-dependent switched systems can be described by i) switching surface and the subspace partitioned by them, ii) subsystem operating in the subspace, and iii) switching law.

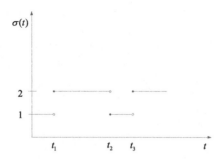

FIGURE 1.2 Switching signal depends on time.

In the time-dependent switched systems, the switching signal is a function of time where switching happens at some specific time instant. For example, if we consider to employ a differential equation to describe a continuous-time switched system: $\dot{x}(t) = f_{\sigma(t)}(x(t))$, where $\sigma(t) : [0, \infty) \to \Omega = \{1, 2, \ldots, m\}$ represents the switching signal, which is usually assumed as right continuous, i.e., $\sigma(t) = \lim_{\tau \to t^+} \sigma(\tau)$. The switching signal $\sigma(t)$ is given in Fig. 1.2, where t_1, t_2, and t_3 represent the switching instants.

2. Autonomous switching and controlled switching: Autonomous switching means that the switching is not regulated by manual control. For example, the switching signal is unknown, and the sudden switching subjects to environmental changes. For example, the networked control systems with data packet dropping can be viewed as a class of switched systems, and the switching law is determined by the packet dropping [70]. In the communication network, the packet dropping cannot be controlled by people. Therefore, this kind of systems is usually named as autonomous switched systems. Vice versa, controlled switching means the switching subjects to the manual control.

The stability of the switched system is not only determined by subsystems but is also dependent on the switching laws. Generally, for a switched system that only contains two subsystems, we have the following conclusions:

1. For a switched system with stable subsystems, if there are no constraints on the switching laws, i.e., the system is under arbitrary switching, then the system can be unstable.
2. For a switched system with unstable subsystems, there can exist a suitable switching law making the system stable.

Therefore, there are *two* basic problems of the stability, which need to be addressed:

Prob. 1 Searching for a stability criteria under arbitrary switching.

Prob. 2 Searching for a specific switching law if the switched system cannot be stable under arbitrary switching.

A sufficient condition for the stability of a switched system in problem 1 is that a common LF should exist for all subsystems. Therefore, solving problem 1 can be transformed into searching the existence condition for the common LF to be constructed. References [71,72] developed some effective methods.

As mentioned before, not all switched systems can be stable under arbitrary switching. Therefore, we need to find some suitable switching laws to regulate the switched system to be stable under these specific switching laws. For problem 2, let us consider switched systems with stable subsystems. If the switching is slow enough, that is, the time interval between two consecutive switching times is long enough, then the switched system can be guaranteed to be stable. The time interval between two consecutive switching times is called DT, which means the time that the switched system dwells on the subsystem. Generally, this class of systems is named the slowly switched systems. A key problem to obtain the stability criterion for such systems is to search for the lower bound of the DT.

It is a very conservative requirement for switched system to have a *fixed* time interval τ such that the DT is always larger than τ. Let us consider a slow switched system with two subsystems A_1 and A_2, in which system performs better on A_1 than on A_2. When the system dwells on A_1 with interval τ and then switches to A_2, this results in a poor performance. Moreover, it takes the system at least time interval τ to switch back to A_1. This results in a switched system with poor performance or, even worse, in an unstable system. A more flexible strategy is to allow the DT between different time points to be, occasionally, less than τ, but on average, larger than τ. This allows the switched system to dwell longer on A_1 than on A_2. This strategy, the so-called *average dwell time* (ADT) ensures not only the stability but also helps to improve the performance of the system [58]. The ADT approach is extensively used to the analysis and synthesis of slow switched systems.

1.1.2.3 Markovian Jump Systems

Since the foundation work developed by Krasovskii and Lidskiid in 1961 [73], the Markovian jump system (MJS) has become a hot topic in the control community, and its research has covered almost every aspect of control theory. It often builds models with a number of actual systems with different working modes, such as solar plants, economic systems, biochemical systems, and network control systems [74,75]. Such systems may run with external environmental changes, actuator fault, communication time delay, data packet loss, and so on. These random factors often cause jumping (or switching) phenomenon of system structure or parameters. One typical example is a solar boiler system. The reflector reflects sunlight onto a high tower, and the boiler on the tower automatically adjusts the flow of water according to the received solar energy. In this control system, different weather conditions, such as sunny days, rainy days, and cloudy days can be regarded as different operating modes, which are often considered as discrete-time events. In each operating mode, the receiving and flow of solar energy have continuous-time dynamics. The different modes can be approximated as a random non-hysteresis-Markov process. This is a typical MJS. Another example is networked control system [85]. Due to the existence of the communication network, the sensor to the controller and the controller to the actuator channels have network-induced delays and packet losses, which often appear with stochastic characteristics. The Markov chain can be used to model the stochastic phenomenon induced by network.

We can conclude that the dynamic behavior of MJS consists of two forms: one is the mode of discrete change, and the Markov chain is described by a set of values in a finite integer set. The other is the state of continuous change described by differential (or difference) equations within each mode. In this sense, the MJS belongs to the category of the hybrid systems, and it particularity contains discrete event and continuous variables, which can unify two different dynamics into stochastic differential (difference) equations. This provides a new way for people to use the state space method in modern control theory to study the analysis and synthesis of MJS. In addition, MJS is also closely related to the switched system [76,77], and each operating mode can be viewed as a subsystem of the whole system. The difference is that the switching law of the MJS is not a human-designed active control law that relies on state or other system parameters, but rather a random process that follows a certain statistical rule. In recent years, the theoretical framework of MJS continuously keeps improving and maturing. On the other hand, many challenging problems need to be solved urgently. This section discusses the stability, robust control, filtering, time-delay, and so on.

The first is the study of system stability. Stability is the most basic requirement, and it is meaningless to discuss other properties unless system stability is ensured. Because of the existence of random factors, regulating MJS to be asymptotically stable is very difficult. Therefore, based on the LF approach, sufficient conditions for the mean square stability and almost sure stability were well studied. Furthermore, by employing the Kronecker product approach, necessary and sufficient conditions for the mean square stability were developed [78,79]. The stability problem for MJS is thoroughly studied by [80]. It is proved that the concept of mean square stability, stochastic stability, and exponential mean square stability are equivalent and they all can be boiled down to sufficient conditions of almost sure uniform stability. Based on the study of the stability analysis, the stabilization problem of MJS also made progress at the same time [81,82]. In MJSs, jumping transition probability (TP) or jumping rate (JR) has important impact on system stability and other performances. In view of system parameters and jumping TP with polytopic uncertainties, the robust stabilization problem for a class of MJSs has been well studied [83]. At the same time, the norm-bounded uncertainties are also considered [84]. Recently, a new research tendency is to study a more general case of MJS with partly unknown TPs or JRs [85–87]. It covers the MJS with completely unknown TPs or JRs and switched system under arbitrary switching.

Many theoretical results of the modern control theory are based on the accurate mathematical model of controlled object. Once the model is biased against the actual object, the controlled system is hard to achieve the desired performance requirements. Based on the inherent inaccuracy of the model-based control method, the robust control theory was developed. It mainly reflects the attenuation capability of the control system to the external disturbance and tolerance of the uncertainty of the model. With the improvement of the robust control theory of linear systems, the related problems of MJS have attracted the attention of many scholars. In the early years, Shi et al. [88] used a Riccati equation to study the H_∞ control problem of uncertain discrete-time MJSs. The robust analysis and synthesis of the underlying system can be converted into feasibility of a Riccati matrix equation. Such an approach was easy for theoretical analysis, but some parameters must be determined in advance. The choice of parameters not only affected the solvability but also the solutions, and it was very difficult to search for optimal values. However, with the development of LMI technique, H_2/H_∞ control has become popular with burgeoning research interests. The design method based on LMI gives convex constraints for the solvability of the optimization problem to meet the design requirements of getting control gain. Costa et al. [89,90] studied the H_2 control and H_2/H_∞ control of MJSs via convex optimization method, and the controller design was boiled down to solving a series of coupling LMIs. Subsequently, in view of the discrete-time case, Costa and Marques [91] revealed the internal relations between the H_2 norm, controllability, and observability and studied the robust H_2 control problem for the MJSs consisting of time-varying uncertainties. Farias et al. [92] aimed at the controller design for MJS with norm bounded uncertainties in both system and input matrices, and Seiler and Sengupta [93] further gave the bounded real lemma, which was more general. Fragoso and Costa [94] gave the sufficient and necessary conditions for the stochastic stabilization of MJS with uncertain jumping parameters. Zhang et al. [95] and Liu et al. [96] gave sufficient conditions of the mean square stability for the MJS with known or unknown jumping knowledge, and the latter gave the design of saturated controller.

Then is the filtering problem. Many comprehensive problems in the control study, including stabilization, pole assignment, asymptotic tracking, and disturbance rejection depend on the convenience of state feedback control. However, in engineering applications, the real state of the system often cannot be directly measured. To solve physical difficulty of the state feedback, an effective method is using the system input and output information to reconstruct state and replace the real state. For the deterministic system, it is usually named the design of state observer, and for the random system, it is often called the filtering problem. If the disturbance in the system or measurement is white noise or the noise with known spectral density, then an optimal filter can be designed by minimizing the variance of the estimated error. Such filters are often in the form of recursive derivations, such as the famous Kalman filter [97,98]. When the statistics of exogenous disturbances are difficult to obtain, it is usually assumed that the disturbance has limited energy and the H_∞ norm of the transfer function, which is from disturbance input to the estimated error, is formulated to achieve H_∞ filtering with an optimized attenuation level [99,100]. Shi et al. [101] and Mahmoud et al. [102] studied the robust Kalman filtering problem of continuous- and discrete-time uncertain MJSs, respectively. They effectively reduced the influence of uncertain factors on the filtering error. Wang et al. [103] studied the filtering problem of a continuous-time nonlinear system with Markov jumping parameters and designed a nonlinear full-order filter. Then, the design method of the robust filter in the discrete-time case was given in the form of a Riccati matrix inequality [104]. In recent years, based on the LMI technique, the robust H_∞ filtering for MJSs got rapid development. Shi, Wu, et al. [105, 106] aimed at MJS with mode-dependent time-delay to discuss the robust performance of the filter and soon extended H_∞ filtering theory to the 2D scenario. In recent years, Liu et al. [107] further developed H_∞ filtering for MJS with unknown jumping knowledge. Moreover, the $L_2 - L_\infty$ filtering and peak–peak filtering has also been studied [108,109].

Finally, there is the study of the time delay MJSs. The existence of time delay can damage the expected performance of the system to some extent and even cause the system instability. The stability criterion for a time delay system generally has two kinds: delay-independent criterion and delay-dependent criterion. An effective approach to solve the stability and robustness of the uncertain time-delay system is based on the Lyapunov stability theory. According to different types of time delay, such as constant time delay [110–112], time-varying delay [113–115], and interval time delay [116–118], it is flexible and useful to select an appropriate LF to search for feasible controllers. In recent years, the LF method for stability analysis and controller synthesis of uncertain time-delay MJSs can be converted into the solutions of LMI, and it greatly promotes the research of the time-delay MJSs. Chen et al. [119] presented a delay-dependent stability criterion when considering mode-dependent time delay and gave the corresponding design method of H_∞ controller. Then, Zhao [120] studied the stability, controllability, and optimal regulation of the time-delay MJS for the influence of external disturbances. Shao [121] aimed at MJS with random noise, model uncertainty, and mode-dependent time delay and gave an H_∞ filtering method that depends on the time-delay interval. Wang et al. [122] studied the descending-order H_∞ filtering on the basis of [121]. Fei et al. [123] further improved the design method of the stabilizing controller by selecting an appropriate Lyapunov–Krasovskii function. Zhao and Zeng [124] presented a delay-dependent analysis method for H_∞ performance. Kang et al. [125] gave an estimation method of time delay for bilinear MJS with actuator saturation. Wu et al. [126,127] designed an H_∞ filter for the general time-delay MJS and gave the robust exponential stability criterion.

However, it is worth mentioning that the NLF approach cannot be directly employed to MJSs with time-invariant TPs because necessary and sufficient condition has already been given by fully making use of the TPs knowledge. Therefore, in this book, we mainly aim at nonhomogenous MJSs subject to time-varying TPs. A typical example is a networked system with packet dropouts and network delays modeled by Markov processes [128,129]. This is due to the fact that delays and packet dropouts are different according to different time periods. Therefore, JRs may be time varying through the working region, and they are usually uncertain. Another example is the helicopter system [130], where the airspeed variations in such system matrices are ideally modeled by a homogeneous Markov chain. However, the TPs of these multiple airspeeds are not fixed when the weather changes. There are similar phenomena in other practical problems, such as robotic manipulators [131], teleoperators [132], and wheeled mobile manipulator systems [133]. In such situations, it is reasonable to model the system as an MJS with nonhomogeneous jump process (chain), that is, the TPs are time varying. One feasible assumption is to use a polytope set to describe uncertainties existing in time-varying TPs. The main reason is that although the TP of the Markov process is not exactly known, we can evaluate their values in some operating points.

1.2 OUTLINE

The research programs and technical routes of the book are shown in Fig. 1.3. The outline of the book is as follows.

Chapter 1 introduces the research background, motivations, research problems (involving robust H_∞ design for T-S fuzzy models, switched systems, and nonhomogenous MJSs using the NLF approach) and concludes by presenting the outline of the book.

Chapter 2 addresses the robust H_∞ state feedback controller design of discrete-time T-S Fuzzy systems by using the NLF approach. A T-S fuzzy

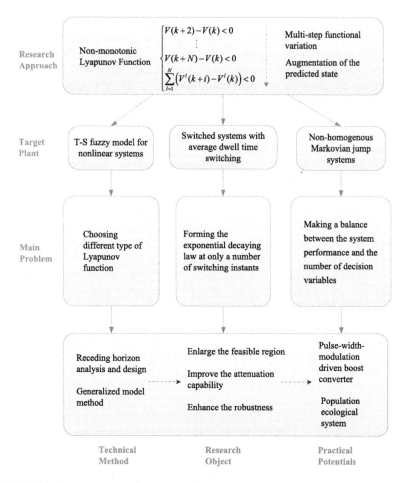

FIGURE 1.3 Research programmes and technical routes.

model of the nonlinear system is often constructed as a combination of linear models with nonlinear scalar membership functions. This motivates researchers to design filters and controllers for nonlinear systems represented by T-S fuzzy systems using the LMI framework. However, many issues related to using LMI fuzzy control, like conservatism, are still not completely solved. This chapter uses an NLF to design a robust H_∞ state feedback controller for uncertain T-S fuzzy systems. In the NLF approach, the monotonicity requirement of the LF is relaxed by allowing it to increase locally but go to zero ultimately. Based on the NLF approach, sufficient conditions for the existence of a robust state feedback H_∞ controller that guarantees stability and a prescribed H_∞ performance are given in terms of LMI. The developed design technique is shown to be less conservative than the existing k-sample variations of LF.

Chapter 3 develops robust H_∞ filtering of T-S Fuzzy systems via NLF. Filtering is an important research issue, particularly in signal processing applications, as it provides means to estimate state information in the presence of disturbance and system uncertainties. In the literature, H_∞ filtering approach has been one of the most popular approaches since it does not require exact knowledge of the disturbance. As discussed in Chapter 2, using NLF successfully reduces the conservatism associated with using an LMI fuzzy control. This chapter adopts the idea of nonmonotonic approach to design a robust H_∞ filter for a T-S fuzzy model of nonlinear systems. Our main contribution is to further reduce the conservatism and improving the H_∞ performance using an NLF.

In Chapter 4, robust H_∞ output feedback control for T-S fuzzy systems is studied via the NLF approach. Motivated by the results in Chapter 2, this chapter develops a robust H_∞ output feedback control stabilization for uncertain T-S fuzzy systems via LMIs. To reduce the conservatism associated with designing output feedback controller for a T-S fuzzy system, a new form of NLF is applied on matrix convexification. Based on the NLF approach, sufficient conditions for the existence of robust H_∞ output feedback control are derived. For comparison, the design technique is shown to be less conservative than the existing nonmonotonic approach, namely, k-sample variations of LF [13].

Chapter 5 concerns the problems of stability and H_∞ control for a class of discrete-time switched systems by introducing a one-step ahead LF approach. The one-step ahead LF is a function of future states. The design objectives reducing the conservatism of the stability criterion developed for arbitrarily switched systems and further getting a better disturbance attenuation capability. The distinguishing feature is that the one-step ahead LF has no structural constraint, such as diagonal structure, and the resulting analysis and synthesis criteria can cover the nonmonotonic method considering two-sample variation, i.e., $V_{k+2}^2 - V_k^2 + V_{k+1}^1 - V_k^1 < 0$ or $V_{k+2} - V_k < 0$ as a particular case.

Chapter 6 investigates an N-step ahead LF approach, which allows a nonmonotonic behavior both at the switching instants and during the running time of each subsystem but guarantees an average decrease at every N sampling steps. The asymptotic stability criterion is improved as well as the capability of disturbance attenuation. By introducing a series of auxiliary variables, the future knowledge of states and exogenous noises can be properly used to derive sufficient conditions for the existence of a robust H_∞ controller in the form of a set of numerical testable conditions. Note that N has a direct impact on the number of the inequality constraints. The essential difficulty is constructing an exponential damping law of the decreasing points of LF, i.e., finding the joint point between the switching interval and the predictive horizon. Moreover, the relationship between the N-step time difference of LF and switching rate, i.e., the ADT constraint, is thoroughly discussed. An ecology system is employed to demonstrate practical potentials of the presented design framework.

In Chapter 7, a nonmonotonic Lyapunov function approach with an N-step ahead predictive horizon is developed to design a robust H_∞ filter for a discrete-time uncertain switched system. By increasing the number N, the filtering performance can be improved as well as the capability of disturbance attenuation. However, the average dwell time (ADT) constraint of the switching law should be more critical as a cost at the same time. To further relax the restriction on the switching law, the mode-dependent ADT switching is introduced to reduce the ADT bound so that a trade-off between the switching frequency and filtering performance can be achieved. Therefore, a codesign of the filter and switching law can be obtained by the developed approach.

Chapter 8 develops dissipative dynamic output feedback (DOF) control for a class of average dwell-time switching system via the multistep Lyapunov function (MLF) approach, which is a relaxed version of NLF. First, a larger dissipative region with guaranteed stability and, specifically, smaller H_∞ level can be achieved by increasing a predictive step N, which means that the monotonic requirement of LF is relaxed. Then, based on the results of dissipative analysis, a robust dissipative DOF controller is further designed. Unlike the traditional method that introduces equality constraint to obtain numerical testable conditions with heuristic nature, a less conservative controller is designed, where the LF matrix is formulated without structural constraint.

In Chapter 9, the robust H_∞ control problem for a class of discrete-time nonhomogenous Markovian jump linear systems (NMJLSs) are investigated by an MLF approach. The multistep LF is allowed to increase during the period of several sampling time step ahead of the current time within the jump mode. First, a less conservative stability criterion is derived based on this multistep LF approach. Second, an H_∞ performance is analyzed under the multistep case by properly dealing with the knowledge of future states and extraneous noises. These two results are then employed to facilitate a robust H_∞ control design for NMJLSs, which yields a better disturbance attenuation performance as compared with the existing results.

Chapter 10 studies the robust H_∞ filtering problem for a class of nonhomogeneous Markovian jump delay systems (NMJDSs) with an N-step ahead Lyapunov–Krasovskii function (NALKF) approach. The N-step ahead is utilized to reduce the conservatism of robust H_∞ filtering. We aim to design filters such that, for all possible time-varying transition probabilities, and all admissible parameter uncertainties and time delays, the filtering error system is mean-square stable with a smaller estimated error and a lower dissipative level. In terms of linear matrix inequalities (LMIs), sufficient conditions for the solvability of the addressed problem are developed via a moving horizon method, i.e., to avoid essential difficulties introduced by future noise, a set of LMIs are obtained at each step of the predictive horizon rather than constructing a huge LMI over the whole horizon.

In Chapter 11, we sum up the results of the book and discuss some related topics for future research work.

References

[1] O. Ameur, P. Massioni, G. Scorletti, X. Brun, M. Smaoui, Lyapunov stability analysis of switching controllers in presence of sliding modes and parametric uncertainties with application to pneumatic systems, IEEE Transactions on Control Systems Technology 24 (6) (2016) 1953–1964.

[2] L.X. Zhang, H.J. Gao, Asynchronously switched control of switched linear systems with average dwell time, Automatica 46 (5) (2010) 953–958.

[3] L.X. Zhang, T. Yang, P. Colaneri, Stability and stabilization of semi-Markov jump linear systems with exponentially modulated periodic distributions of sojourn time, IEEE Transactions on Automatic Control 62 (6) (2017) 2870–2885.

[4] M. Sato, Gain-scheduled output-feedback controllers depending solely on scheduling parameters via parameter-dependent Lyapunov functions, Automatica 47 (12) (2011) 2786–2790.

[5] F. de Oliveira Souza, M.C. de Oliveira, R.M. Palhares, A simple necessary and sufficient LMI condition for the strong delay-independent stability of LTI systems with single delay, Automatica 89 (2018) 407–410.

[6] A.R. Butz, Higher order derivatives of Liapunov functions, IEEE Transactions on Automatic Control 14 (1) (1969) 111–112.

[7] J.A. Yorke, A theorem on Liapunov functions using $\ddot{V}(k) < 0$, Mathematical Systems Theory 4 (1970) 40–45.

[8] D. Aeyels, J. Peuteman, A new asymptotic stability criterion for nonlinear time-variant differential equations, IEEE Transactions on Automatic Control 43 (7) (1998) 968–971.

[9] A.A. Ahmadi, P.A. Parrilo, Non-monotonic Lyapunov functions for stability of discrete time nonlinear and switched systems, in: Proceedings of the 47th IEEE Conference on Decision and Control, 2008, pp. 614–621.

[10] S.F. Derakhshan, A. Fatehi, Non-monotonic Lyapunov functions for stability analysis and stabilization of discrete time Takagi–Sugeno fuzzy systems, International Journal of Innovative Computing, Information & Control (2014) 1567–1586.

[11] S.F. Derakhshan, A. Fatehi, M.G. Sharabiany, Nonmonotonic observer-based fuzzy controller designs for discrete time T-S fuzzy systems via LMI, IEEE Transactions on Cybernetics 44 (12) (2014) 2557–2567.

[12] S.F. Derakhshan, A. Fatehi, Non-monotonic robust H_2 fuzzy observer-based control for discrete time nonlinear systems with parametric uncertainties, International Journal of Systems Science 46 (12) (2015) 2134–2149.

[13] A. Kruszewski, R. Wang, T.M. Guerra, Nonquadratic stabilization conditions for a class of uncertain nonlinear discrete time T-S fuzzy models: a new approach, IEEE Transactions on Automatic Control 53 (2) (2008) 606–611.

[14] A. Nasiri, S.K. Nguang, A. Swain, D.J. Almakhles, Reducing conservatism in H_∞ robust state feedback control design of T-S fuzzy systems: a non-monotonic approach, IEEE Transactions on Fuzzy Systems 26 (1) (2018) 386–390.

[15] Y.J. Chen, M. Tanaka, K. Inoue, H. Ohtake, et al., A nonmonotonically decreasing relaxation approach of Lyapunov functions to guaranteed cost control for discrete fuzzy systems, IET Control Theory & Applications 8 (16) (2014) 1716–1722.

[16] A. Nasiri, S.K. Nguang, A. Swain, D.J. Almakhles, Robust output feedback controller design of discrete-time Takagi–Sugeno fuzzy systems: a non-monotonic Lyapunov approach, IET Control Theory & Applications 10 (5) (2016) 545–553.

[17] K. Tanaka, H.O. Wang, Fuzzy Control Systems Design and Analysis: A Linear Matrix Inequality Approach, Wiley, 2001.

[18] W. Pedrycz, M. Reformat, Rule-based modeling of nonlinear relationships, IEEE Transactions on Fuzzy Systems 5 (2) (1997) 256–269.

[19] C. Fantuzzi, R. Rovatti, On the approximation capacities of the homogeneous Takagi–Sugeno model, in: Proceedings of IEEE 5th International Fuzzy Systems, 1996, pp. 1067–1072.

[20] T. Takagi, M. Sugeno, Fuzzy identification of systems and its applications to modeling and control, IEEE Transactions on Systems, Man and Cybernetics SMC-15 (1) (1985) 116–132.

[21] D. Yue, Q.L. Hang, C. Peng, State feedback controller design of networked control systems, IEEE Transactions on Circuits and Systems. II, Express Briefs 51 (11) (2004) 640–644.

[22] H.O. Wang, K. Tanaka, M.F. Griffin, An approach to fuzzy control of nonlinear systems: stability and design issues, IEEE Transactions on Fuzzy Systems 4 (1) (1996) 14–23.

[23] K. Tanaka, H.O. Wang, Fuzzy Control Systems Design and Analysis: A Linear Matrix Inequality Approach, Wiley, 2001.

[24] J. Dong, G.H. Yang, Static output feedback control synthesis for discrete-time T-S fuzzy systems, International Journal of Control, Automation, and Systems 5 (3) (2007) 349–354.

[25] K. Tanaka, T. Ikeda, H.O. Wang, Fuzzy regulators and fuzzy observers: relaxed stability conditions and LMI-based designs, IEEE Transactions on Fuzzy Systems 6 (2) (1998) 250–265.

[26] K. Tanaka, T. Ikeda, H.O. Wang, Robust stabilization of a class of uncertain nonlinear systems via fuzzy control: quadratic stabilizability, H_∞ control theory and linear matrix inequalities, IEEE Transactions on Fuzzy Systems 4 (1) (1996) 1–13.

[27] S.K. Nguang, P. Shi, H_∞ fuzzy output feedback control design for nonlinear systems: an LMI approach, IEEE Transactions on Fuzzy Systems 11 (3) (2003) 331–340.

[28] W. Assawinchaichote, S.K. Nguang, P. Shi, E.K. Boukas, H_∞ fuzzy state-feedback control design for nonlinear systems with stability constraints: an LMI approach, Mathematics and Computers in Simulation 78 (4) (2008) 514–531.

[29] D. Huang, S.K. Nguang, State feedback control of uncertain networked control systems with random time delays, IEEE Transactions on Automatic Control 53 (3) (2008) 829–834.

[30] S. Nguang, W. Assawinchaichote, H_∞ filtering for fuzzy dynamical systems with D-stability constraints, IEEE Transactions on Circuits and Systems. I, Fundamental Theory and Applications 50 (11) (2003) 1503–1508.

[31] S. Nguang, P. Shi, Delay-dependent H_∞ filtering for uncertain time delay nonlinear systems: an LMI approach, IET Control Theory & Applications 1 (1) (2007) 133–140.

[32] P. Shi, Y. Zhang, M. Chadli, R. Agarwal, Mixed H_∞ and passive filtering for discrete fuzzy neural networks with stochastic jumps and time delays, IEEE Transactions on Neural Networks and Learning Systems 27 (4) (2016) 903–909.

[33] B. Jiang, Z. Mao, P. Shi, H_∞ filter design for a class of networked control systems via T-S fuzzy-model approach, IEEE Transactions on Fuzzy Systems 18 (1) (2010) 201–208.

[34] H. Li, H. Liu, H. Gao, P. Shi, Reliable fuzzy control for active suspension systems with actuator delay and fault, IEEE Transactions on Fuzzy Systems 20 (2) (2012) 342–357.

[35] B. Jiang, K. Zhang, P. Shi, Integrated fault estimation and accommodation design for discrete-time Takagi–Sugeno fuzzy systems with actuator faults, IEEE Transactions on Fuzzy Systems 19 (2) (2011) 291–304.

[36] Z.G. Wu, P. Shi, H. Su, J. Chu, Reliable H_∞ control for discrete-time fuzzy systems with infinite-distributed delay, IEEE Transactions on Fuzzy Systems 20 (1) (2012) 22–31.

[37] L. Wang, H. Shi, An LMI approach to robust guaranteed cost fault-tolerant control for a class of uncertain parameter systems, in: 7th World Congress on Intelligent Control and Automation, 2008, pp. 955–959.

[38] K. Tanaka, H. Ohtake, H. Wang, Guaranteed cost control of polynomial fuzzy systems via a sum of squares approach, IEEE Transactions on Systems, Man and Cybernetics 39 (2) (2009) 561–567.

[39] N. Sofianos, O. Kosmidou, Guaranteed cost LMI-based fuzzy controller design for discrete-time nonlinear systems with polytopic uncertainties, in: 18th Mediterranean Conference on Control Automation, 2010, pp. 1383–1388.

[40] Y.J. Chen, M. Tanaka, K. Inoue, H. Ohtake, K. Tanaka, T. Guerra, A. Kruszewski, H. Wang, A nonmonotonically decreasing relaxation approach of Lyapunov functions to guaranteed cost control for discrete fuzzy systems, IET Control Theory & Applications 8 (16) (2014) 1716–1722.

[41] G. Feng, Stability analysis of discrete-time fuzzy dynamic systems based on piecewise Lyapunov functions, IEEE Transactions on Fuzzy Systems 12 (1) (2004) 22–28.

[42] W.J. Wang, Y.J. Chen, C.H. Sun, Relaxed stabilization criteria for discrete-time T-S fuzzy control systems based on a switching fuzzy model and piecewise Lyapunov function, IEEE Transactions on Systems, Man and Cybernetics. Part B. Cybernetics 37 (3) (2007) 551–559.

[43] H. Lam, Polynomial fuzzy-model-based control systems: stability analysis via piecewise-linear membership functions, IEEE Transactions on Fuzzy Systems 19 (3) (2011) 588–593.

[44] M. Johansson, A. Rantzer, K. Arzen, Piecewise quadratic stability of fuzzy systems, IEEE Transactions on Fuzzy Systems 7 (6) (1999) 713–722.

[45] Y. Chen, M. Tanaka, K. Tanaka, H. Wang, Stability analysis and region-of-attraction estimation using piecewise polynomial Lyapunov functions: polynomial fuzzy model approach, IEEE Transactions on Fuzzy Systems 23 (4) (2015) 1314–1322.

[46] M. Johansson, A. Rantzer, K. Arzen, Piecewise quadratic stability of fuzzy systems, IEEE Transactions on Fuzzy Systems 7 (6) (1999) 713–722.

[47] K. Tanaka, T. Hori, H. Wang, A multiple Lyapunov function approach to stabilization of fuzzy control systems, IEEE Transactions on Fuzzy Systems 11 (4) (2003) 582–589.

[48] G.J. Liu, Y.Y. Cao, J.L. Chen, A new approach to fault detection observer design of nonlinear time-delay systems via fuzzy Lyapunov functions, in: 31st Chinese Control Conference, 2012, pp. 951–956.

[49] B.J. Rhee, S. Won, A new fuzzy Lyapunov function approach for a Takagi–Sugeno fuzzy control system design, Fuzzy Sets and Systems 157 (9) (2006) 1211–1228.

[50] K. Tanaka, T. Hori, H. Wang, New parallel distributed compensation using time derivative of membership functions: a fuzzy Lyapunov approach, in: Proceedings of the 40th IEEE Conference on Decision and Control, Orland, 2001, pp. 3942–3947.

[51] J. Li, S. Zhou, S. Xu, Fuzzy control system design via fuzzy Lyapunov functions, IEEE Transactions on Systems, Man and Cybernetics. Part B. Cybernetics 38 (6) (2008) 1657–1661.

[52] H.K. Lam, Stability analysis of T-S fuzzy control systems using parameter-dependent Lyapunov function, IET Control Theory & Applications 3 (6) (2009) 750–762.

[53] S. Chae, S.K. Nguang, SoS based robust H_∞ fuzzy dynamic output feedback control of nonlinear networked control systems, IEEE Transactions on Cybernetics 44 (7) (2014) 1204–1213.

[54] K. Guelton, N. Manamanni, C.C. Duong, D.L. Koumba-Emianiwe, Sum-of-squares stability analysis of Takagi–Sugeno systems based on multiple polynomial Lyapunov functions, International Journal of Fuzzy Systems 15 (1) (2013) 1–8.

[55] C. Perez, V. Azhmyakov, A. Poznyak, Practical stabilization of a class of switched systems: dwell-time approach, IMA Journal of Mathematical Control and Information 32 (4) (2015) 689–702.

[56] L.B. Wu, H. Wang, Asynchronous adaptive fault-tolerant control for a class of switched nonlinear systems with mode-dependent dwell time, IEEE Access 5 (2017) 22092–22100.

[57] M.A. Bagherzadeh, J. Askari, J. Ghaisari, M. Mojiri, Robust asymptotic stability of parametric switched linear systems with dwell time, IET Control Theory & Applications 12 (4) (2018) 477–483.

[58] J. Hespanha, A. Morse, Stability of switched systems with average dwell-time, in: Proceedings of the 38th IEEE Conference on Decision Control, Phoenix, AZ, 1999, pp. 2655–2660.

[59] Y.G. Chen, S.M. Fei, K.J. Zhang, Stability analysis for discrete-time switched linear singular systems: average dwell time approach, IMA Journal of Mathematical Control and Information 30 (2) (2013) 239–249.

[60] N. Li, J.D. Cao, Switched exponential state estimation and robust stability for interval neural networks with the average dwell time, IMA Journal of Mathematical Control and Information 32 (2) (2015) 257–276.

[61] D.H. Zheng, H.B. Zhang, Q.X. Zheng, Consensus analysis of multi-agent systems under switching topologies by a topology-dependent average dwell time approach, IET Control Theory & Applications 11 (3) (2017) 429–438.

[62] Y. Wu, Y.P. Wu, Mode-dependent robust stability and stabilisation of uncertain networked control systems via an average dwell time switching approach, IET Control Theory & Applications 11 (11) (2017) 1726–1735.

[63] L.X. Zhang, Y.S. Leng, P. Colaneri, Stability and stabilization of discrete-time semi-Markov jump linear systems via semi-Markov kernel approach, IEEE Transactions on Automatic Control 61 (2) (2016) 503–508.

[64] L.X. Zhang, T. Yang, P. Shi, Y.Z. Zhu, Analysis and Design of Markov Jump Systems with Complex Transition Probabilities, Springer, 2016.

[65] L.X. Zhang, T. Yang, P. Shi, P. Colaneri, Stability and stabilization of semi-Markov jump linear systems with exponentially modulated periodic distributions of sojourn time, IEEE Transactions on Automatic Control 62 (6) (2017) 2870–2885.

[66] L.X. Zhang, H.J. Gao, Asynchronously switched control of switched linear systems with average dwell time, Automatica 46 (5) (2010) 953–958.

[67] L.X. Zhang, N.G. Cui, M. Liu, Y. Zhao, Asynchronous filtering of discrete-time switched linear systems with average dwell time, IEEE Transactions on Circuits and Systems I: Regular Papers 58 (5) (2011) 1109–1118.

[68] L.X. Zhang, Y.Z. Zhu, P. Shi, Y.X. Zhao, Resilient asynchronous H_∞ filtering for Markov jump neural networks with unideal measurements and multiplicative noises, IEEE Transactions on Cybernetics 45 (12) (2015) 2840–2852.

[69] A.V. der Schaft, J. Schumacher, An Introduction to Hybrid Dynamical Systems, Springer, New York, 2000.

[70] W.A. Zhang, L. Yu, Output feedback stabilization of networked control systems with packet dropouts, IEEE Transactions on Automatic Control 52 (9) (2007) 1705–1710.

[71] Z. Sun, S.S. Ge, Switched Linear Systems: Control and Design, Springer, London, 2005.

[72] D. Liberzon, Switching in Systems and Control, Birkhäuser, Boston, 2003.

[73] N.N. Krasovskii, E.A. Lidskii, Analysis design of controller in systems with random attributes, Part 1, Automatic and Remote Control 22 (1961) 1021–1025.

[74] E.K. Boukas, Stochastic Switching Systems: Analysis and Design, Birkhäuser, Basel, Berlin, 2005.

[75] O.L.V. Costa, M.D. Fragoso, R.P. Marques, Discrete Time Markovian Jump Linear Systems, Springer, London, 2005.

[76] Z.D. Sun, Stabilization and optimization of switched linear systems, Automatica 42 (2006) 783–788.

[77] V.F. Montagner, V.J.S. Leite, R.C.L.F. Oliveira, P.L.D. Peres, State feedback control of switched linear systems, Journal of Computational and Applied Mathematics 194 (2) (2006) 192–206.

[78] M. Mariton, Jump linear quadratic control with random state discontinuities, Automatica 23 (2) (1987) 237–240.

[79] M. Mariton, Detection delays false alarm rates and the reconfiguration of control systems, International Journal of Control 49 (3) (1989) 981–992.

[80] X. Feng, K.A. Loparo, Y. Ji, H.J. Chizeck, Stochastic stability properties of jump linear systems, IEEE Transactions on Automatic Control 37 (1) (1992) 38–53.

[81] Y. Ji, H.J. Chizeck, Controllability, stabilizability, and continuous-time Markovian jump linear quadratic control, IEEE Transactions on Automatic Control 35 (7) (1990) 777–788.

[82] O.L.V. Costa, M.D. Fragoso, Stability results for discrete-time linear systems with Markovian jumping parameters, Journal of Mathematical Analysis and Applications 179 (1) (1993) 154–178.

[83] J.L. Xiong, J. Lam, H.J. Gao, W.C. Daniel, On robust stabilization of Markovian jump systems with uncertain switching probabilities, Automatica 41 (5) (2005) 897–903.

[84] M. Karan, P. Shi, C.Y. Kaya, Transition probability bounds for the stochastic stability robustness of continuous- and discrete-time Markovian jump linear systems, Automatica 42 (12) (2006) 2159–2168.

[85] L.X. Zhang, E.K. Boukas, Stability and stabilization of Markovian jump linear systems with partly unknown transition probabilities, Automatica 45 (2) (2009) 463–468.

[86] L.X. Zhang, E.K. Boukas, Mode-dependent H_∞ filtering for discrete-time Markovian jump linear systems with partly unknown transition probabilities, Automatica 45 (6) (2009) 1462–1467.

[87] G.L. Wang, Q.L. Zhang, V. Sreeram, Partially mode-dependent H_∞ filtering for Markovian jump systems with partly unknown transition probabilities, Signal Processing 90 (2) (2010) 548–556.

[88] P. Shi, E.K. Boukas, R.K. Agarwal, Robust control for Markovian jumping discrete-time systems, International Journal of Systems Science 30 (8) (1999) 787–797.

[89] O.L.V. Costa, J.B.R. Val, J.C. Geromel, A convex programming approach to H2 control of discrete-time Markovian jump linear systems, International Journal of Control 66 (4) (1997) 557–579.

[90] O.L.V. Costa, R.P. Marques, Mixed H_2/H_∞ control of discrete-time Markovian linear systems, IEEE Transactions on Automatic Control 43 (1) (1998) 95–100.

[91] O.L.V. Costa, R.P. Marques, Robust H_2-control for discrete-time Markovian jump linear systems, International Journal of Control 73 (1) (2000) 11–21.

[92] D.P. Farias, J.C. Geromel, J.B.R. Val, A note on the robust control of Markov jump linear uncertain systems, Optimal Control Applications & Methods 23 (2) (2002) 105–112.

[93] P. Seiler, R. Sengupta, A bounded real lemma for jump systems, IEEE Transactions on Automatic Control 48 (9) (2003) 1651–1654.

[94] M.D. Fragoso, O.L.V. Costa, Mean square stabilizability of continuous-time linear systems with partial information on the Markovian jumping parameters, in: American Control Conference, vol. 6, Chicago, 2000, pp. 4299–4303.

[95] L.J. Zhang, C.W. Li, D.Z. Chen, Robust adaptive control of Markov jump systems with parameter uncertainties, Control and Decision 20 (9) (2005) 1030–1037 (in Chinese).

[96] H.P. Liu, E.K. Boukas, F. Sun, D.W.C. Ho, Controller design for Markovian jumping systems subject to actuator saturation, Automatica 42 (3) (2006) 459–465.

[97] R.E. Kalman, A new approach to linear filtering and predication problem, Journal of Basic Engineering 82 (1) (1960) 35–45.

[98] R.E. Kalman, New result in linear filtering and predication theory, Journal of Basic Engineering 83 (1961) 95–108.

[99] S.K. Nguang, P. Shi, H_∞ filtering of nonlinear sampled-data systems, Automatica 36 (2) (2000) 303–310.

[100] L. Xie, C.E. Souza, Y. Wang, Robust filtering for a class of discrete-time uncertain nonlinear systems: an H_∞ approach, International Journal of Robust and Nonlinear Control 6 (4) (1996) 297–312.

[101] P. Shi, E.K. Boukas, R.K. Agarwal, Kalman filtering for continuous-time uncertain systems with Markovian jumping parameters, IEEE Transactions on Automatic Control 44 (8) (1999) 1592–1597.

[102] M.S. Mahmoud, P. Shi, A. Ismail, Robust Kalman filtering for discrete-time Markovian jump systems with parameter uncertainty, Journal of Computational and Applied Mathematics 169 (1) (2004) 53–69.

[103] Z.D. Wang, J. Lam, X.H. Liu, Nonlinear filtering for state delayed systems with Markovian switching, IEEE Transactions on Automatic Control 51 (9) (2003) 2321–2328.

[104] Z.D. Wang, J. Lam, X.H. Liu, Robust filtering for discrete-time Markovian jump delay systems, IEEE Signal Processing Letters 11 (8) (2004) 659–662.

[105] P. Shi, M.S. Mahmoud, S.K. Nguang, A. Ismal, Robust filtering for jumping systems with mode-dependent delays, Signal Processing 86 (1) (2006) 140–152.

[106] L.G. Wu, P. Shi, H.J. Gao, C.H. Wang, H_∞ filtering for 2D Markovian jump system, Automatica 44 (7) (2008) 1848–1858.

[107] H.P. Liu, D.W.C. Ho, F.C. Sun, Design of H_∞ filter for Markov jumping linear systems with non-accessible mode information, Automatica 44 (10) (2008) 2655–2660.

[108] F. Liu, Robust $L_2 - L_\infty$ filtering for uncertain jump systems, Control and Decision 20 (1) (2005) 32–35.

[109] S.P. He, F. Liu, Robust peak-to-peak filtering for Markov jump systems, Signal Processing 90 (2) (2010) 513–522.

[110] M.S. Ali, Z.K. Hou, M.N. Noori, Stability and performance of feedback control systems with time delays, Computers & Structures 66 (2–3) (1998) 241–248.

[111] P.L. Liu, Exponential stability for linear time-delay systems with delay dependence, Journal of the Franklin Institute 340 (6–7) (2003) 481–488.

[112] S.Y. Xu, J. Lam, T.W. Chen, Robust H_∞ control for uncertain discrete stochastic time-delay systems, Systems & Control Letters 5 (3–4) (2004) 203–215.

[113] S.H. Song, J.K. Kim, C.H. Yim, H.C. Kim, H_∞ control of discrete-time linear systems with time-varying delays in state, Automatica 35 (9) (1999) 1587–1591.

[114] S.Y. Xu, J. Lam, Y. Zou, New results on delay-dependent robust H_∞ control for systems with time-varying delays, Automatica 42 (2) (2006) 343–348.

[115] F. Liu, M. Wu, Y. He, R. Yokoyama, New delay-dependent stability criteria for T-S fuzzy systems with time-varying delay, Fuzzy Sets and Systems 161 (5) (2010) 2033–2042.

[116] X.F. Jiang, Q.L. Han, Delay-dependent robust stability for uncertain linear systems with interval time-varying delay, Automatica 42 (6) (2006) 1059–1065.

[117] X.F. Jiang, Q.L. Han, New stability criteria for linear systems with interval time-varying delay, Automatica 44 (10) (2008) 2680–2685.

[118] Y. Zhang, Y. He, M. Wu, Delay-dependent robust stability for uncertain stochastic systems with interval time-varying delay, Acta Automatica Sinica 35 (5) (2009) 577–582.

[119] W.H. Chen, Z.H. Guan, P. Yu, Delay-dependent stability and H_∞ control of uncertain discrete-time Markovian jump systems with mode-dependent time delays, Systems & Control Letters 52 (5) (2004) 361–376.

[120] P. Zhao, Practical stability, controllability and optimal control of stochastic Markovian jump systems with time-delays, Automatica 44 (12) (2008) 3120–3125.

[121] H.Y. Shao, Delay-range-dependent robust H_∞ filtering for uncertain stochastic systems with mode-dependent time delays and Markovian jump parameters, Journal of Mathematical Analysis and Applications 342 (2) (2008) 1084–1095.

[122] G.L. Wang, Q.L. Zhang, V. Sreeram, Design of reduced-order H_∞ filtering for Markovian jump systems with mode-dependent time delays, Signal Processing 89 (2) (2009) 187–196.

[123] Z.Y. Fei, H.J. Gao, P. Shi, New results on stabilization of Markovian jump systems with time delay, Automatica 45 (10) (2009) 2300–2306.

[124] X.D. Zhao, Q.S. Zeng, New robust delay-dependent stability and H_∞ analysis for uncertain Markovian jump systems with time-varying delays, Journal of the Franklin Institute 347 (5) (2010) 863–874.

[125] Y. Kang, W.K. Shang, H.S. Xi, Estimating the delay-time for the stability of Markovian jump bilinear systems with saturating actuators, Acta Automatica Sinica 36 (5) (2010) 762–767.

[126] Z.G. Wu, H.Y. Su, J. Chu, Delay-dependent H_∞ filtering for singular Markovian jump time-delay systems, Signal Processing 90 (6) (2010) 1815–1824.

[127] Z.G. Wu, H.Y. Su, J. Chu, Robust exponential stability of uncertain singular Markovian jump time-delay systems, Acta Automatica Sinica 36 (4) (2010) 558–563.

[128] R. Krtolica, et al., Stability of linear feedback systems with random communication delays, International Journal of Control 59 (4) (1994) 925–953.

[129] P. Seiler, R. Sengupta, An H_∞ approach to networked control, IEEE Transactions on Automatic Control 50 (3) (2005) 356–364.

[130] K.S. Narendra, S.S. Tripathi, Identification and optimization of aircraft dynamics, Journal of Aircraft 10 (4) (1973) 193–199.

[131] Y. Kang, Z. Li, X. Cao, D. Zhai, Robust control of motion/force for robotic manipulators with random time delays, IEEE Transactions on Control Systems Technology 21 (5) (2013) 1708–1718.

[132] Z. Li, L. Ding, H. Gao, G. Duan, C. Su, Trilateral teleoperation of adaptive fuzzy force/motion control for nonlinear teleoperators with communication random delays, IEEE Transactions on Fuzzy Systems 21 (4) (2013) 610–623.

[133] Y. Kang, Z. Li, Y. Dong, H. Xi, Markovian based fault-tolerant control for wheeled mobile manipulators, IEEE Transactions on Control Systems Technology 20 (1) (2012) 266–276.

Robust H_∞ State Feedback Controller Design of Discrete-Time T-S Fuzzy Systems: A Nonmonotonic Approach

2.1 INTRODUCTION

Over the past 20 years, Takagi–Sugeno (T-S) fuzzy models have been successfully utilized to approximate nonlinear systems [1]. A T-S fuzzy model of the system is constructed as a combination of linear models with nonlinear scalar membership functions. Such a feature inspires many researchers to use this model for analysis and synthesis of a large class of nonlinear control systems. The stability analysis of a T-S model and subsequently the controller design are often carried out through the standard Lyapunov method. This inherently leads to some inequalities in which the nonlinear membership functions are excluded from final analysis and synthesis conditions. As a result, the stability conditions, which are often constructed in the form of linear matrix inequalities (LMIs) [2], become only sufficient (conservative).

Initially, researchers have considered classic quadratic Lyapunov functions to design controllers for discrete-time T-S fuzzy systems (see [3], [4], and references therein). In many cases, using the standard common quadratic LF yields conservative results [5,6]. Therefore, extensive works have been presented in the literature to mitigate the demerits of this conservatism. For example, various types of Lyapunov functions have been presented, such as piecewise [7], polynomial [8], fuzzy LFs [9], and combinations of these. It is worth noting that type-2 fuzzy systems open new doors in handling uncertainties and nonlinearities, and some remarkable results have been reported in this area, such as [10] and [11]. Another idea to reduce the conservatism in discrete-time systems is relaxing the monotonic decrease of LF (alternately referred to as the nonmonotonic approach) [12–18]. In other words, the LF is not required to decrease in each successive step (i.e., $V_{t+1} < V_t$) [12]. In recent years, many researchers have used the nonmonotonic approach in controller design. For example, in [13, 15,14], the nonmonotonic decrease of LF is based on only 2-sample variations (i.e., $V_{t+2} < V_t$). This nonmonotonic approach has been shown to be less conservative than the common quadratic LF [19] and even piecewise LF [20]. A more general nonmonotonic Lyapunov function (NLF) with k-sample variations (i.e., $V_{t+k} < V_t$) has been introduced in [16–18]. In [16] and [17], a robust state feedback controller for uncertain discrete-time T-S fuzzy systems has been proposed. Moreover, the same approach is considered for the guaranteed cost control design of discrete T-S fuzzy systems in [18].

Motivated by the aforementioned progress, we derive new stability conditions for discrete-time T-S fuzzy systems using a new NLF approach, which further reduces the conservatism. In our method, more relaxed conditions are derived to cover previous studies [13,15,14,16–18] as particular cases. Thus, the first objective of this chapter is to provide new stability conditions for discrete-time T-S fuzzy systems using a new NLF approach. Then, an H_∞ robust state-feedback controller, which is often designed to minimize the ratio of the controlled output energy to the disturbance energy [21], is investigated for uncertain T-S fuzzy systems. To the best knowledge of the authors, the design of an H_∞ controller for uncertain T-S fuzzy systems has not been reported yet using an NLF approach. Finally, sufficient conditions for the existence of an H_∞ controller are derived in terms of LMIs.

Notations: $(\cdot)^T$ and $\bullet$ denote, respectively, the transpose and the symmetric block in a symmetric matrix. The identity matrix of size n is denoted by I_n, and 0_{nm} represents the zero matrix of size $n \times m$; $x_t \triangleq x(t)$ and $\mathbb{K}_{r_0,r} \triangleq \{r_0, \ldots, r\}$, where $t, r_0, \ldots, r \in \mathbb{N}$; $X(v_t) \triangleq \sum_{i=1}^{r} \mu_i(v_t) X_i$ with $\mu_i(v_t) \geq 0$ and $\sum_{i=1}^{r} \mu_i(v_t) = 1$.

2.2 SYSTEM DESCRIPTION AND PROBLEM FORMULATION

This chapter considers the following uncertain discrete-time T-S fuzzy system:

Plant Rule i: IF $v_1(t)$ is M_1^i AND $\cdots$ AND $v_\varrho(t)$ is M_ϱ^i, THEN

$$
\begin{aligned}
x_{t+1} &= (A_i + \Delta A_i)\, x_t + \left(B_{1,i} + \Delta B_{1,i}\right) u_t + D_i w_t, \\
z_t &= C_i x_t + B_{2,i} u_t,
\end{aligned}
\tag{2.1}
$$

where $x_t \in \mathbb{R}^n$, $u_t \in \mathbb{R}^m$, $z_t \in \mathbb{R}^p$, and $w_t \in \mathbb{R}^d$ are the state, control, output signals, and disturbance, respectively, $i \in \mathbb{K}_{1,r}$ denotes the ith fuzzy inference rule, r is the number of rules, $\upsilon_\mathfrak{g}(t)$ are the premise variables ($\upsilon_t = [\upsilon_1(t), \ldots, \upsilon_\varrho(t)]$), ϱ is the number of premise variables, and $M_1^i, \ldots, M_\varrho^i$ are membership functions. The matrices A_i, ΔA_i, $B_{1,i}$, $\Delta B_{1,i}$, D_i, C_i, and $B_{2,i}$ are of appropriate dimensions. By using a center-average defuzzifier, product inference, and singleton fuzzifier, the uncertain T-S fuzzy system (2.1) can be expressed as follows:

$$x_{t+1} = (A(\upsilon_t) + \Delta A(\upsilon_t))x_t + (B_1(\upsilon_t) + \Delta B_1(\upsilon_t))u_t + D(\upsilon_t)w_t,$$

$$z_t = C(\upsilon_t)x_t + B_2(\upsilon_t)u_t, \tag{2.2}$$

where $\mu_i(\upsilon_t) = \dfrac{\prod_{\mathfrak{g}=1}^{\varrho} M_\mathfrak{g}^i(\upsilon_\mathfrak{g}(t))}{\sum_{\ell=1}^{r} \prod_{\mathfrak{g}=1}^{\varrho} M_\mathfrak{g}^\ell(\upsilon_\mathfrak{g}(t))}$ with $\mu_i(\upsilon_t) \geq 0$ and $\sum_{i=1}^{r} \mu_i(\upsilon_t) = 1$. The system uncertainties are assumed to be $[\Delta A(\upsilon_t) \ \Delta B_1(\upsilon_t)] = HF(x_t, t)[E^a(\upsilon_t) \ E^b(\upsilon_t)]$, where H, $E^a(\upsilon_t)$, and $E^b(\upsilon_t)$ are known matrix functions, which characterize the structure of the uncertainties, and $F(x_t, t)$ is an unknown nonlinear time-varying matrix function satisfying $F(x_t, t)^T F(x_t, t) \leq I$.

This chapter solves the following nonparallel distributed controller (PDC) state feedback H_∞ problem.

H_∞ *State Feedback Problem Formulation:* Given a prescribed H_∞ performance γ, design a non-PDC state feedback controller

$$u_t = K(\upsilon_t)x_t, \tag{2.3}$$

where $K(\upsilon_t) = Y(\upsilon_t)G^{-1}(\upsilon_t)$ is such that system (2.1) with controller (2.3) is stable and the following inequality holds:

$$\sum_{t=0}^{\infty} z_t^T z_t < \gamma^2 \sum_{t=0}^{\infty} \left(w_t^T w_t \right). \tag{2.4}$$

The T-S fuzzy system (2.2) with the fuzzy controller (2.3) gives the following closed loop system:

$$x_{t+1} = \hat{A}(\upsilon_t)x_t + D(\upsilon_t)w_t,$$

$$z_t = \hat{C}(\upsilon_t)x_t, \tag{2.5}$$

where $\hat{A}(\upsilon_t) = A(\upsilon_t) + B_1(\upsilon_t)K(\upsilon_t) + \Delta A(\upsilon_t) + \Delta B_1(\upsilon_t)K(\upsilon_t)$ and $\hat{C}(\upsilon_t) = C(\upsilon_t) + B_2(\upsilon_t)K(\upsilon_t)$.

Before proceeding to the next section, let us recall the following lemmas.

Lemma 2.1. *[23] Suppose that the following conditions are satisfied for all combinations of $c_l \in \mathbb{K}_{1,r}$ and $l \in \mathbb{K}_{1,k}$:*

$$\begin{cases} \Lambda_{ii}^{c_1...c_k} < 0 & \forall\, i \in \mathbb{K}_{1,r}, \\ \frac{1}{r-1}\Lambda_{ii}^{c_1...c_k} + \frac{1}{2}\left(\Lambda_{ij}^{c_1...c_k} + \Lambda_{ij}^{c_1...c_k}\right) < 0 & \forall\, 1 < i \neq j < r. \end{cases} \quad (2.6)$$

Then we have the following parameterized LMI:

$$\sum_{c_k=1}^{r} \cdots \sum_{c_1=1}^{r} \sum_{j=1}^{r} \sum_{i=1}^{r} \mu_i(v_t)\mu_j(v_t)\mu_{c_1}(v_{t+1})...\mu_{c_k}(v_{t+k})\Lambda_{ij}^{c_1...c_k} < 0. \quad (2.7)$$

Lemma 2.2. *[12] Consider a discrete-time nonlinear dynamic system $x_{t+1} = f(x_t)$. Suppose that there exist functions $V_1(x_t), \ldots, V_k(x_t) : \mathbb{R}^n \to \mathbb{R}$ such that*

$$\sum_{i=1}^{k} i V_i(0) = 0, \ \sum_{i=j}^{k} V_i(x_t) > 0 \ \forall\, x_t \neq 0, \ j \in \mathbb{K}_{1,k}, \text{ and radially unbounded,}$$

(2.8a)

$$\Delta_k V(x_t) = \sum_{i=1}^{k} (V_i(x_{t+i}) - V_i(x_t)) < 0.$$

(2.8b)

Then the origin is a globally asymptotically stable equilibrium of the system.

Lemma 2.3. *[22] For a symmetric matrix $P > 0$ and a matrix G, we have $G^T + G - P \leq G^T P^{-1} G$.*

Remark 2.1. Lemma 2.2 states that as long as the sum of k improvements is negative at any point in time, the stability is guaranteed. If $V_i = 0$ for $i = 1, \ldots, k - 1$ in Lemma 2.2, then conditions (2.8a)–(2.8b) are the same conditions as in [13–18], that is, $V_k(x_t) > 0$ and $V_k(x_{t+k}) < V_k(x_t)$.

2.3 MAIN RESULTS

In this section, we present first stability conditions for the T-S fuzzy systems using the NLF approach. Then we give sufficient conditions for the existence of an H_∞ robust state feedback control for uncertain discrete-time T-S fuzzy systems.

2.3.1 Stability Analysis

In this subsection, we consider system (2.1) with $\Delta A(v_t) = \Delta B_1(v_t) = w_t = u_t = 0$, that is,

$$x_{t+1} = A(v_t)x_t. \tag{2.9}$$

The following theorem gives stability conditions for system (2.9).

Theorem 2.1. *The T-S fuzzy system (2.9) is globally asymptotically stable if there exists a set of symmetric matrices P_{i,i_0} and $Q_{i,i_0,...,i_k}$ for $i \in \mathbb{K}_{1,k}$ and $i_0, i_1, ..., i_k \in \mathbb{K}_{1,r}$ such that*

$$\sum_{i=j}^{k} P_{i,i_0} > 0 \qquad \forall\, j \in \mathbb{K}_{1,k} \text{ and } i_0 \in \mathbb{K}_{1,r}, \tag{2.10}$$

$$A_{i_{k-1}}^T P_{k,i_k} A_{i_{k-1}} + P_{k-1,i_{k-1}} < Q_{k-1,i_0...i_{k-1}}, \tag{2.11a}$$

$$A_{i_{k-s}}^T Q_{k-s+1,i_0...i_{k-s+1}} A_{i_{k-s}} + P_{k-s,i_{k-s}} < Q_{k-s,i_0...i_{k-s}} \,\, \forall\, s \in \mathbb{K}_{2,k-1}, \tag{2.11b}$$

$$A_{i_0}^T Q_{1,i_0 i_1} A_{i_0} < \sum_{i=1}^{k} P_{i,i_0}. \tag{2.11c}$$

Proof. Consider the function

$$V_i(x_t) \triangleq x_t^T P_i(v_t)x_t \quad \forall\, i \in \mathbb{K}_{1,k}, \tag{2.12}$$

where

$$P_i(v_t) = \sum_{i_0=1}^{r} \mu_{i_0}(v_t) P_{i,i_0}. \tag{2.13}$$

Since $\mu_{i_0}(v_t) \geq 0$ and $\sum_{i_0=1}^{r} \mu_{i_0}(v_t) = 1$ for all $i_0 \in \mathbb{K}_{1,r}$, if (2.10) holds, then

$$\sum_{i=j}^{k} P_i(v_t) > 0 \,\, \forall\, j \in \mathbb{K}_{1,k}. \tag{2.14}$$

The following steps prove that

$$\Delta_k V(x_t) = \sum_{i=1}^{k} \left(x_{t+i}^T P_i(v_{t+i})x_{t+i} - x_t^T P_i(v_t)x_t \right) \tag{2.15}$$

is less than zero, which in turn establishes the global asymptotic stability of (2.9) according to Lemma 2.2. To simplify our notation, define

$$\Pi_L \triangleq \sum_{i_L=1}^{r} \cdots \sum_{i_0=1}^{r} \mu_{i_0}(v_t)...\mu_{i_L}(v_{t+L})Q_{L,i_0,..i_L} \quad \forall\, L \in \mathbb{K}_{1,k-1}. \tag{2.16}$$

Step 1: Adding and subtracting $x_{t+k-1}^T \Pi_{k-1} x_{t+k-1}$ to and from (2.15), we have

$$
\begin{aligned}
\Delta_k V(x_t) \quad &= x_{t+k}^T P_k(v_{t+k}) x_{t+k} + x_{t+k-1}^T P_{k-1}(v_{t+k-1}) x_{t+k-1} \\
&\quad - x_{t+k-1}^T \Pi_{k-1} x_{t+k-1} + x_{t+k-1}^T \Pi_{k-1} x_{t+k-1} \\
&\quad + \sum_{i=1}^{k-2} x_{t+i}^T P_i(v_{t+i}) x_{t+i} - \sum_{i=1}^{k} x_t^T P_i(v_t) x_t.
\end{aligned}
\tag{2.17}
$$

Using the system equation (2.9), we have $x_{t+k} = A(v_{t+k-1}) x_{t+k-1}$ and

$$
\begin{aligned}
\Delta_k V(x_t) \quad &= x_{t+k-1}^T \big(A^T(v_{t+k-1}) P_k(v_{t+k}) A(v_{t+k-1}) \\
&\quad + P_{k-1}(v_{t+k-1}) - \Pi_{k-1} \big) x_{t+k-1} \\
&\quad + x_{t+k-1}^T \Pi_{k-1} x_{t+k-1} + \sum_{i=1}^{k-2} x_{t+i}^T P_i(v_{t+i}) x_{t+i} \\
&\quad - \sum_{i=1}^{k} x_t^T P_i(v_t) x_t.
\end{aligned}
\tag{2.18}
$$

If (2.11a) holds, then $A^T(v_{t+k-1}) P_k(v_{t+k}) A(v_{t+k-1}) + P_{k-1}(v_{t+k-1}) - \Pi_{k-1} < 0$, which leads to

$$
\Delta_k V(x_t) < x_{t+k-1}^T \Pi_{k-1} x_{t+k-1} + \sum_{i=1}^{k-2} x_{t+i}^T P_i(v_{t+i}) x_{t+i} - \sum_{i=1}^{k} x_t^T P_i(v_t) x_t.
\tag{2.19}
$$

Step 2: Adding and subtracting $x_{t+k-2}^T \Pi_{k-2} x_{t+k-2}$ to and from (2.19), we have

$$
\begin{aligned}
\Delta_k V(x_t) \quad &< x_{t+k-1}^T \Pi_{k-1} x_{t+k-1} + x_{t+k-2}^T P_{k-2}(v_{t+k-2}) x_{t+k-2} \\
&\quad - x_{t+k-2}^T \Pi_{k-2} x_{t+k-2} + x_{t+k-2}^T \Pi_{k-2} x_{t+k-2} \\
&\quad + \sum_{i=1}^{k-3} x_{t+i}^T P_i(v_{t+i}) x_{t+i} - \sum_{i=1}^{k} x_t^T P_i(v_t) x_t.
\end{aligned}
\tag{2.20}
$$

Using the system equation (2.9), we have $x_{t+k-1} = A(v_{t+k-2}) x_{t+k-2}$ and

$$
\begin{aligned}
\Delta_k V(x_t) \quad &< x_{t+k-2}^T \big(A^T(v_{t+k-2}) \Pi_{k-1} A(v_{t+k-2}) \\
&\quad + P_{k-2}(v_{t+k-2}) - \Pi_{k-2} \big) x_{t+k-2} + x_{t+k-2}^T \Pi_{k-2} x_{t+k-2} \\
&\quad + \sum_{i=1}^{k-3} x_{t+i}^T P_i(v_{t+i}) x_{t+i} - \sum_{i=1}^{k} x_t^T P_i(v_t) x_t.
\end{aligned}
\tag{2.21}
$$

If (2.11b) holds for $s = 2$, then $A^T(v_{t+k-2})\Pi_{k-1}A(v_{t+k-2}) + P_{k-2}(v_{t+k-2}) - \Pi_{k-2} < 0$, which leads to

$$\Delta_k V(x_t) < x_{t+k-2}^T \Pi_{k-2} x_{t+k-2} + \sum_{i=1}^{k-3} x_{t+i}^T P_i(v_{t+i})x_{t+i} - \sum_{i=1}^{k} x_t^T P_i(v_t)x_t.$$

(2.22)

Repeating the procedure $k - 3$ times gives

$$\Delta_k V(x_t) < x_{t+1}^T \Pi_1 x_{t+1} - \sum_{i=1}^{k} x_t^T P_i(v_t)x_t.$$

(2.23)

Step k: Using (2.9), we obtain

$$\Delta_k V(x_t) < x_t^T \left(A^T(v_t)\Pi_1 A(v_t) - \sum_{i=1}^{k} P_i(v_t) \right) x_t.$$

(2.24)

If (2.11c) holds, then $A^T(v_t)\Pi_1 A(v_t) - \sum_{i=1}^{k} P_i(v_t) < 0$ and $\Delta_k V(x_t) < 0$. By Lemma 2.2 system (2.9) is globally asymptotically stable. This completes the proof. □

In the following, we adopt the example from [17] to illustrate that the developed approach in this chapter is less conservative in comparison with existing results [17].

Example 2.1. Consider the system matrices $A_1 = \begin{bmatrix} a_2 & -0.5 \\ a_1 & -0.9 \end{bmatrix}$ and $A_2 = \begin{bmatrix} -0.9 & -a_1 \\ -0.6 & 0.8 \end{bmatrix}$, where a_1 and a_2 are variables. Fig. 2.1 shows the stability regions for $k = 2$ and $k = 3$ obtained using Theorem 1 in [17] and Theorem 2.1. The points marked with "o" and "×" show the stability region obtained using Theorem 2.1 and Theorem 1 in [17], respectively. It is obvious from Fig. 2.1 that the developed approach yields larger stability regions compared with existing results.

2.3.2 H_∞ Robust Control Design

The following theorem provides sufficient conditions for the existence of the H_∞ robust state feedback control for the uncertain system (2.5) using an NLF approach.

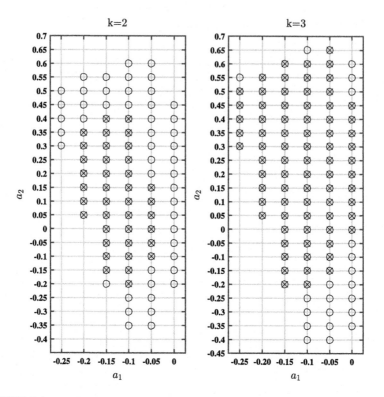

FIGURE 2.1 Stability region: ○ – Theorem 2.1 and × – Theorem 1 [17].

Theorem 2.2. *Given a constant $\gamma > 0$, the closed-loop control system (2.5) is globally asymptotically stable with disturbance attenuation γ if there exist positive scalar parameters ρ_i, a set of symmetric matrices P_{i,i_0}, $Q_{L,i_0,j_0...,i_L j_L}$, and matrices $G_{i_\mathfrak{h}}$, $Y_{j_\mathfrak{h}}$, where $\mathfrak{h} \in \mathbb{K}_{0,k-1}$, $i_0...i_k j_0...j_{k-1} \in \mathbb{K}_{1,r}$, $L \in \mathbb{K}_{1,k-1}$, and $i \in \mathbb{K}_{1,k}$, such that*

$$\sum_{i=j}^{k} P_{i,i_0} > 0 \qquad \forall\, i_0 \in \mathbb{K}_{1,r} \text{ and } j \in \mathbb{K}_{1,k}, \tag{2.25}$$

$$\begin{cases} \Lambda_{1,i_{k-1}i_{k-1}} < 0, \\ \frac{1}{r-1}\Lambda_{1,i_{k-1}i_{k-1}} + \frac{1}{2}\left(\Lambda_{1,i_{k-1}j_{k-1}} + \Lambda_{1,j_{k-1}i_{k-1}}\right) < 0 \,\forall\, i_{k-1} \neq j_{k-1}, \end{cases} \tag{2.26a}$$

$$\begin{cases} \Lambda_{s,i_{k-s}i_{k-s}} < 0, \\ \frac{1}{r-1}\Lambda_{s,i_{k-s}i_{k-s}} + \frac{1}{2}\left(\Lambda_{s,i_{k-s}j_{k-s}} + \Lambda_{s,j_{k-s}i_{k-s}}\right) < 0 \,\forall\, i_{k-s} \neq j_{k-s}, \end{cases}$$

$$\forall\, s \in \mathbb{K}_{2,k-1}, \tag{2.26b}$$

$$\begin{cases} \Lambda_{k,i_0 i_0} < 0, \\ \frac{1}{r-1}\Lambda_{k,i_0 i_0} + \frac{1}{2}\left(\Lambda_{k,i_0 j_0} + \Lambda_{k,j_0 i_0}\right) < 0 \ \forall \ i_0 \neq j_0, \end{cases} \tag{2.26c}$$

where

$$\Lambda_{1,i_{k-1}j_{k-1}} = \begin{bmatrix} \begin{pmatrix} P_{k-1,i_{k-1}} \\ -Q_{k-1,i_0 j_0..i_{k-1}j_{k-1}} \end{pmatrix} & \bullet & \bullet & \bullet & \bullet & \bullet \\ 0_{dn} & -\gamma I_d & \bullet & \bullet & \bullet & \bullet \\ \bar{C}_{i_{k-1}j_{k-1}} & 0_{pd} & -\gamma I_p & \bullet & \bullet & \bullet \\ \check{A}_{i_{k-1}j_{k-1}} & D_{i_{k-1}} & 0_{np} & -G_{i_k}^T - G_{i_k} + P_{k,i_k} & \bullet & \bullet \\ \bar{E}_{i_{k-1}j_{k-1}} & 0_{nd} & 0_{np} & 0_{nn} & -\rho_1 I_n & \bullet \\ 0_{nn} & 0_{nd} & 0_{np} & H^T & 0_{nn} & -\rho_1^{-1} I_n \end{bmatrix}, \tag{2.27}$$

$$\Lambda_{s,i_{k-s}j_{k-s}} = \begin{bmatrix} \begin{pmatrix} P_{k-s,i_{k-s}} \\ -Q_{k-s,i_0 j_0..i_{k-s}j_{k-s}} \end{pmatrix} & \bullet & \bullet & \bullet & \bullet & \bullet \\ 0_{dn} & -\gamma I_d & \bullet & \bullet & \bullet & \bullet \\ \bar{C}_{i_{k-s}j_{k-s}} & 0_{pd} & -\gamma I_p & \bullet & \bullet & \bullet \\ \check{A}_{i_{k-s}j_{k-s}} & D_{i_{k-s}} & 0_{np} & \begin{pmatrix} -G_{i_{k-s+1}}^T - G_{i_{k-s+1}} \\ +Q_{k-s+1,i_0 j_0..i_{k-s+1}j_{k-s+1}} \end{pmatrix} & \bullet & \bullet \\ \bar{E}_{i_{k-s}j_{k-s}} & 0_{nd} & 0_{np} & 0_{nn} & -\rho_s I_n & \bullet \\ 0_{nn} & 0_{nd} & 0_{np} & H^T & 0 & -\rho_s^{-1} I_n \end{bmatrix}, \tag{2.28}$$

$$\Lambda_{k,i_0 j_0} = \begin{bmatrix} -\sum_{i=1}^{k} P_{i,i_0} & \bullet & \bullet & \bullet & \bullet & \bullet \\ 0_{dn} & -\gamma I_d & \bullet & \bullet & \bullet & \bullet \\ \bar{C}_{i_0 j_0} & 0_{pd} & -\gamma I_p & \bullet & \bullet & \bullet \\ \check{A}_{i_0 j_0} & D_{i_0} & 0_{np} & -G_{i_1}^T - G_{i_1} + Q_{1,i_0 j_0 i_1 j_1} & \bullet & \bullet \\ \bar{E}_{i_0 j_0} & 0_{nd} & 0_{np} & 0_{nn} & -\rho_k I_n & \bullet \\ 0_{nn} & 0_{nd} & 0_{np} & H^T & 0_{nn} & -\rho_k^{-1} I_n \end{bmatrix}, \tag{2.29}$$

$$\bar{C}_{i_\mathfrak{h} j_\mathfrak{h}} = C_{i_\mathfrak{h}} G_{j_\mathfrak{h}} + B_{2,i_\mathfrak{h}} Y_{j_\mathfrak{h}}, \tag{2.30}$$

$$\check{A}_{i_\mathfrak{h} j_\mathfrak{h}} = A_{i_\mathfrak{h}} G_{j_\mathfrak{h}} + B_{1,i_\mathfrak{h}} Y_{j_\mathfrak{h}}, \tag{2.31}$$

$$\bar{E}_{i_\mathfrak{h} j_\mathfrak{h}} = E_{i_\mathfrak{h}}^a G_{j_\mathfrak{h}} + E_{i_\mathfrak{h}}^b Y_{j_\mathfrak{h}}. \tag{2.32}$$

Proof. Consider the function

$$\hat{V}_i(x_t) \triangleq x_t^T \hat{P}_i(v_t)x_t \quad \forall\, i \in \mathbb{K}_{1,k}, \tag{2.33}$$

where

$$\hat{P}_i(v_t) = G^{-T}(v_t)P_i(v_t)G^{-1}(v_t). \tag{2.34}$$

Similarly to the proof of Theorem 2.1, if (2.10) holds, then

$$\sum_{i=j}^{k} \hat{P}_i(v_t) > 0 \quad \forall\, j \in \mathbb{K}_{1,k}, \tag{2.35}$$

which implies that $\sum_{i=j}^{k} \hat{V}_i(x_t) > 0$. It is necessary to check the existence of $G^{-1}(v_t)$. Note that if (2.26a) holds, then $G_{i_k}^T + G_{i_k} - P_{k,i_k} > 0$. This in turn implies that $G^T(v_t) + G(v_t) - P_k(v_t) > 0$. From (2.35), $P_k(v_t)$ is a positive definite matrix, and thus $G^T(v_t) + G(v_t) > P_k(v_t) > 0$. Hence this implies that $G(v_t)$ is nonsingular.

Let us denote

$$\Delta_k \hat{V}(x_t) \triangleq \sum_{i=1}^{k} \left(x_{t+i}^T \hat{P}_i(v_{t+i})x_{t+i} - x_t^T \hat{P}_i(v_t)x_t \right). \tag{2.36}$$

To simplify our notation, let $\hat{\Pi}_L \triangleq G^{-T}(v_{t+L})\Pi_i G^{-1}(v_{t+L})$, where

$$\Pi_L = \sum_{i_L=1}^{r} \sum_{j_L=1}^{r} \cdots \sum_{i_0=1}^{r} \sum_{j_0=1}^{r} \mu_{i_0}(v_t)\mu_{j_0}(v_t)\ldots\mu_{i_L}(v_{t+L})\mu_{j_L}(v_{t+L})Q_{L,i_0 j_0..i_L j_L}$$

and $\quad L \in \mathbb{K}_{1,k-1}.$

Step 1: Adding and subtracting $x_{t+k-1}^T \hat{\Pi}_{k-1} x_{t+k-1} + \frac{1}{\gamma} z_{t+k-1}^T z_{t+k-1} - \gamma w_{t+k-1}^T w_{t+k-1}$ to and from $\Delta_k \hat{V}(x_t)$ defined in (2.15) and rearranging its terms give

$$\begin{aligned}
\Delta_k \hat{V}(x_t) &= x_{t+k}^T \hat{P}_k(v_{t+k})x_{t+k} + x_{t+k-1}^T \hat{P}_{k-1}(v_{t+k-1})x_{t+k-1} \\
&\quad - x_{t+k-1}^T \hat{\Pi}_{k-1}x_{t+k-1} + \frac{1}{\gamma} z_{t+k-1}^T z_{t+k-1} - \gamma w_{t+k-1}^T w_{t+k-1} \\
&\quad + x_{t+k-1}^T \hat{\Pi}_{k-1}x_{t+k-1} + \sum_{i=1}^{k-2} x_{t+i}^T \hat{P}_i(v_{t+i})x_{t+i} \\
&\quad - \frac{1}{\gamma} z_{t+k-1}^T z_{t+k-1} + \gamma w_{t+k-1}^T w_{t+k-1} - x_t^T \left(\sum_{i=1}^{k} \hat{P}_i(v_t) \right) x_t.
\end{aligned} \tag{2.37}$$

Define $\zeta_t^T \triangleq [\ x_t^T \quad w_t^T\ ]$. Using the system equation (2.5), we have $x_{t+k} = \hat{A}(v_{t+k-1})x_{t+k-1} + D(v_{t+k-1})w_{t+k-1}$. Substituting this into (2.37) gives

$$
\begin{aligned}
\Delta_k \hat{V}(x_t) &= \zeta_{t+k-1}^T \chi_1 \zeta_{t+k-1} + x_{t+k-1}^T \hat{\Pi}_{k-1} x_{t+k-1} + \sum_{i=1}^{k-2} x_{t+i}^T \hat{P}_i(v_{t+i}) x_{t+i} \\
&\quad - \frac{1}{\gamma} z_{t+k-1}^T z_{t+k-1} + \gamma w_{t+k-1}^T w_{t+k-1} - x_t^T \left(\sum_{i=1}^k \hat{P}_i(v_t) \right) x_t,
\end{aligned}
\tag{2.38}
$$

where

$$
\chi_1 = \begin{bmatrix} \chi_1^{11} & \chi_1^{12} \\ \bullet & \chi_1^{22} \end{bmatrix},
\tag{2.39}
$$

$$
\begin{aligned}
\chi_1^{11} &= \hat{A}^T(v_{t+k-1})\hat{P}_k(v_{t+k})\hat{A}(v_{t+k-1}) - \hat{\Pi}_{k-1} \\
&\quad + \hat{P}_{k-1}(v_{t+k-1}) + \frac{1}{\gamma}\hat{C}^T(v_{t+k-1})\hat{C}(v_{t+k-1}), \\
\chi_1^{12} &= \hat{A}^T(v_{t+k-1})\hat{P}_k(v_{t+k})D(v_{t+k-1}), \\
\chi_1^{22} &= D^T(v_{t+k-1})\hat{P}_k(v_{t+k})D(v_{t+k-1}) - \gamma I_d.
\end{aligned}
\tag{2.40}
$$

In the rest of this step, we will show that (2.26a) implies that χ_1 is less than zero.

Pre- and postmultiplying (2.39) by $diag\{G^T(v_{t+k-1}), I\}$ and its transpose, respectively, give

$$
\Xi_1 = \begin{bmatrix} \Xi_1^{11} & \Xi_1^{12} \\ \bullet & \chi_1^{22} \end{bmatrix},
\tag{2.41}
$$

$$
\begin{aligned}
\Xi_1^{11} &= \bar{A}^T(v_{t+k-1})\hat{P}_k(v_{t+k})\bar{A}(v_{t+k-1}) \\
&\quad - \Pi_{k-i} + P_{k-i}(v_{t+k-i}) + \frac{1}{\gamma}\bar{C}^T(v_{t+k-1})\bar{C}(v_{t+k-1}), \\
\Xi_1^{12} &= \bar{A}^T(v_{t+k-1})\hat{P}_k(v_{t+k})D(v_{t+k-1}), \\
\bar{A}(v_t) &= \hat{A}(v_t)G(v_t) = \check{A}(v_t) + HF_t\bar{E}(v_t), \\
\check{A}(v_t) &= A(v_t)G(v_t) + B_1(v_t)Y(v_t), \\
\bar{E}(v_t) &= E^a(v_t)G(v_t) + E^b(v_t)Y(v_t), \\
\bar{C}(v_t) &= \hat{C}(v_t)G(v_t) = C(v_t)G(v_t) + B_2(v_t)Y(v_t).
\end{aligned}
\tag{2.42}
$$

Using Schur's complement, we can show that $\varXi_1 < 0$ if

$$
\begin{bmatrix}
-\Pi_{k-1} + P_{k-1}(v_{t+k-1}) & \bullet & \bullet & \bullet \\
0_{dn} & -\gamma I_d & \bullet & \bullet \\
\bar{C}(v_{t+k-1}) & 0_{pd} & -\gamma I_p & \bullet \\
\bar{A}(v_{t+k-1}) & D(v_{t+k-1}) & 0_{np} & -\hat{P}_{t+k}^{-1}
\end{bmatrix} < 0. \qquad (2.43)
$$

Noting that $-\hat{P}_{t+k}^{-1} \le -G^T(v_{t+k}) - G(v_{t+k}) + P_k(v_{t+k})$, we get that (2.43) holds if

$$
\Upsilon_1 = \begin{bmatrix}
-\Pi_{k-1} + P_{k-1}(v_{t+k-1}) & \bullet & \bullet & \bullet \\
0_{dn} & -\gamma I_d & \bullet & \bullet \\
\bar{C}(v_{t+k-1}) & 0_{pd} & -\gamma I_p & \bullet \\
\bar{A}(v_{t+k-1}) & D(v_{t+k-1}) & 0_{np} & \begin{pmatrix} -G^T(v_{t+k}) - G(v_{t+k}) \\ +P_k(v_{t+k}) \end{pmatrix}
\end{bmatrix} < 0.
$$

$$\qquad (2.44)$$

Defining $\quad \tilde{H} \triangleq \begin{bmatrix} 0 & 0 & 0 & 0 \\ 0 & 0 & 0 & 0 \\ 0 & 0 & 0 & 0 \\ H & 0 & 0 & 0 \end{bmatrix}$, $\tilde{F}_{t+k-1} \triangleq \begin{bmatrix} F_{t+k-1} & 0 & 0 & 0 \\ 0 & 0 & 0 & 0 \\ 0 & 0 & 0 & 0 \\ 0 & 0 & 0 & 0 \end{bmatrix},$

$\tilde{E}_{t+k-1} \triangleq \begin{bmatrix} \bar{E}(v_{t+k-1}) & 0 & 0 & 0 \\ 0 & 0 & 0 & 0 \\ 0 & 0 & 0 & 0 \\ 0 & 0 & 0 & 0 \end{bmatrix}$, we get that Υ_1 can be expressed as

$\Upsilon_1 = \Upsilon_0 + \tilde{H}\tilde{F}_{t+k-1}\tilde{E}_{t+k-1} + \left(\tilde{H}\tilde{F}_{t+k-1}\tilde{E}_{t+k-1}\right)^T$, where

$$
\Upsilon_0 \triangleq \begin{bmatrix}
-\Pi_{k-1} + P_{k-1}(v_{t+k-1}) & \bullet & \bullet & \bullet \\
0_{dn} & -\gamma I_d & \bullet & \bullet \\
\bar{C}(v_{t+k-1}) & 0_{pd} & -\gamma I_p & \bullet \\
\check{A}(v_{t+k-1}) & D(v_{t+k-1}) & 0_{np} & \begin{pmatrix} -G^T(v_{t+k}) - G(v_{t+k}) \\ +P_k(v_{t+k}) \end{pmatrix}
\end{bmatrix}. \qquad (2.45)
$$

By the inequality in Lemma 2.3, (2.44) holds if

$$
\begin{bmatrix}
\begin{pmatrix} -\Pi_{k-1} + P_{k-1}(v_{t+k-1}) \\ +\rho_1^{-1}\bar{E}_{t+k-1}^T\bar{E}_{t+k-1} \end{pmatrix} & \bullet & \bullet & \bullet \\
0_{dn} & -\gamma I_d & \bullet & \bullet \\
\bar{C}(v_{t+k-1}) & 0_{pd} & -\gamma I_p & \bullet \\
\check{A}(v_{t+k-1}) & D(v_{t+k-1}) & 0_{np} & \begin{pmatrix} -G^T(v_{t+k}) - G(v_{t+k}) \\ +P_k(v_{t+k}) + \rho_1 HH^T \end{pmatrix}
\end{bmatrix} < 0. \quad (2.46)
$$

Applying Schur's complement to (2.46) gives

$$
\begin{bmatrix}
-\Pi_{k-1} + P_{k-1}(v_{t+k-1}) & \bullet & \bullet & \bullet & \bullet & \bullet \\
0_{dn} & -\gamma I_d & \bullet & \bullet & \bullet & \bullet \\
\bar{C}(v_{t+k-1}) & 0_{pd} & -\gamma I_p & \bullet & \bullet & \bullet \\
\check{A}(v_{t+k-1}) & D(v_{t+k-1}) & 0_{np} & \begin{pmatrix} -G^T(v_{t+k}) - G(v_{t+k}) \\ +P_k(v_{t+k}) \end{pmatrix} & \bullet & \bullet \\
\bar{E}(v_{t+k-1}) & 0_{nd} & 0_{np} & 0_{nn} & -\rho_1 I_n & \bullet \\
0_{nn} & 0_{nd} & 0_{np} & H^T & 0_{nn} & -\rho_1^{-1} I_n
\end{bmatrix} < 0.
$$
$$(2.47)$$

According to Lemma 2.1, if (2.26a) holds, then so does (2.47). Hence it implies that $\chi_1 < 0$. Knowing that $\chi_1 < 0$, from (2.38) we have

$$
\begin{aligned}
\Delta_k \hat{V}(x_t) \quad & < x_{t+k-1}^T \hat{\Pi}_{k-1} x_{t+k-1} + \sum_{i=1}^{k-2} x_{t+i}^T \hat{P}_i(v_{t+i}) x_{t+i} \\
& -\frac{1}{\gamma} z_{t+k-1}^T z_{t+k-1} + \gamma w_{t+k-1}^T w_{t+k-1} - x_t^T \left(\sum_{i=1}^{k} \hat{P}_i(v_t) \right) x_t.
\end{aligned}
\quad (2.48)
$$

Step 2: Adding and subtracting $x_{t+k-2}^T \hat{\Pi}_{k-2} x_{t+k-2} + \frac{1}{\gamma} z_{t+k-2}^T z_{t+k-2} - \gamma w_{t+k-2}^T w_{t+k-2}$ to and from (2.48) and rearranging its terms give

$$
\begin{aligned}
\Delta_k \hat{V}(x_t) \quad & < x_{t+k-1}^T \hat{\Pi}_{k-1} x_{t+k-1} + x_{t+k-2}^T \hat{P}_{k-2}(v_{t+k-2}) x_{t+k-2} \\
& -x_{t+k-2}^T \hat{\Pi}_{k-2} x_{t+k-2} + \frac{1}{\gamma} z_{t+k-2}^T z_{t+k-2} - \gamma w_{t+k-2}^T w_{t+k-2} \\
& +x_{t+k-2}^T \hat{\Pi}_{k-2} x_{t+k-2} + \sum_{i=1}^{k-3} x_{t+i}^T \hat{P}_i(v_{t+i}) x_{t+i} \\
& -\sum_{i=k-2}^{k-1} \left(\frac{1}{\gamma} z_{t+i}^T z_{t+i} - \gamma w_{t+i}^T w_{t+i} \right) - x_t^T \left(\sum_{i=1}^{k} \hat{P}_i(v_t) \right) x_t.
\end{aligned}
\quad (2.49)
$$

Using the system equation (2.5), we have $x_{t+k-1} = \hat{A}(v_{t+k-2})x_{t+k-2} + D(v_{t+k-2})w_{t+k-2}$ and substituting this into (2.49), we obtain

$$
\begin{aligned}
\Delta_k \hat{V}(x_t) \quad &< \zeta_{t+k-2}^T \chi_2 \zeta_{t+k-2} \\
&+ x_{t+k-2}^T \hat{\Pi}_{k-2} x_{t+k-2} + \sum_{i=1}^{k-3} x_{t+i}^T \hat{P}_i(v_{t+i})x_{t+i} \\
&- \sum_{i=k-2}^{k-1} \left(\frac{1}{\gamma} z_{t+i}^T z_{t+i} - \gamma w_{t+i}^T w_{t+i} \right) - x_t^T \left(\sum_{i=1}^{k} \hat{P}_i(v_t) \right) x_t,
\end{aligned} \tag{2.50}
$$

where

$$
\chi_2 = \begin{bmatrix} \chi_2^{11} & \chi_2^{12} \\ \bullet & \chi_2^{22} \end{bmatrix}, \tag{2.51}
$$

$$
\begin{aligned}
\chi_2^{11} &= \hat{A}^T(v_{t+k-2})\hat{\Pi}_{k-1}\hat{A}(v_{t+k-2}) - \hat{\Pi}_{k-2} + \hat{P}_{k-2}(v_{t+k-2}) \\
&\quad + \frac{1}{\gamma}\hat{C}^T(v_{t+k-2})\hat{C}(v_{t+k-2}), \\
\chi_2^{12} &= \hat{A}^T(v_{t+k-2})\hat{\Pi}_{k-1}D(v_{t+k-2}), \\
\chi_2^{22} &= D^T(v_{t+k-2})\hat{\Pi}_{k-1}D(v_{t+k-2}) - \gamma I_d.
\end{aligned} \tag{2.52}
$$

Following the same procedure of *Step 1*, (2.47) implies $\chi_2 < 0$, and we obtain

$$
\begin{aligned}
\Delta_k \hat{V}(x_t) \quad &< x_{t+k-2}^T \hat{\Pi}_{k-2} x_{t+k-2} + \sum_{i=1}^{k-3} x_{t+i}^T \hat{P}_i(v_{t+i})x_{t+i} \\
&- \sum_{i=k-2}^{k-1} \left(\frac{1}{\gamma} z_{t+i}^T z_{t+i} - \gamma w_{t+i}^T w_{t+i} \right) - x_t^T \left(\sum_{i=1}^{k} \hat{P}_i(v_t) \right) x_t.
\end{aligned} \tag{2.53}
$$

Repeating the procedure $k-3$ times, we arrive at

$$
\begin{aligned}
\Delta_k \hat{V}(x_t) \quad &< x_{t+1}^T \hat{\Pi}_1 x_{t+1} - x_t^T \left(\sum_{i=1}^{k} \hat{P}_i(v_t) \right) x_t \\
&- \sum_{i=1}^{k-1} \left(\frac{1}{\gamma} z_{t+i}^T z_{t+i} - \gamma w_{t+i}^T w_{t+i} \right).
\end{aligned} \tag{2.54}
$$

Step k: Adding and subtracting $\frac{1}{\gamma} z_t^T z_t - \gamma w_t^T w_t$ to and from (2.54), using the system equation (2.5) again $x_{t+1} = \hat{A}(v_t)x_t + D(v_t)w_t$, and rearranging its terms give

$$
\Delta_k \hat{V}(x_t) < \zeta_t^T \chi_k \zeta_t - \sum_{i=0}^{k-1} \left(\frac{1}{\gamma} z_{t+i}^T z_{t+i} - \gamma w_{t+i}^T w_{t+i} \right), \tag{2.55}
$$

where

$$
\chi_k = \begin{bmatrix} \chi_k^{11} & \chi_k^{12} \\ \bullet & \chi_k^{22} \end{bmatrix},
\tag{2.56}
$$

$$
\begin{aligned}
\chi_k^{11} &= \hat{A}^T(v_t)\hat{\Pi}_1\hat{A}(v_t) - \sum_{i=1}^{k}\hat{P}_i(v_t) + \frac{1}{\gamma}\hat{C}^T(v_t)\hat{C}(v_t), \\
\chi_k^{12} &= \hat{A}^T(v_t)\hat{\Pi}_1 D(v_t), \\
\chi_k^{22} &= D^T(v_t)\hat{\Pi}_1 D(v_t) - \gamma I_d.
\end{aligned}
\tag{2.57}
$$

Similarly, we can show that (2.26c) implies $\chi_k < 0$. Hence, from (2.55) we have

$$
\Delta_k \hat{V}(x_t) + \sum_{i=0}^{k-1}\left(\frac{1}{\gamma}z_{t+i}^T z_{t+i} - \gamma w_{t+i}^T w_{t+i}\right) < 0.
\tag{2.58}
$$

Taking the summation from $t = -k$ to $t = \infty$ on both sides of (2.58), we have

$$
\sum_{t=-k}^{\infty}\left\{\Delta_k \hat{V}(x_t) + \sum_{i=0}^{k-1}\left(\frac{1}{\gamma}z_{t+i}^T z_{t+i} - \gamma w_{t+i}^T w_{t+i}\right)\right\} < 0.
\tag{2.59}
$$

Substituting (2.16) into (2.59) and expanding summations give

$$
\sum_{j=0}^{k-1}\sum_{i=j}^{k-1}\left(\hat{V}_{i+1}(x_\infty) - \hat{V}_{i+1}(x_{i-k})\right) + \sum_{i=0}^{k-1}\left(\frac{1}{\gamma}\sum_{t=0}^{\infty}z_t^T z_t - \gamma\sum_{t=0}^{\infty}w_t^T w_t\right) < 0.
\tag{2.60}
$$

Under zero initial conditions ($x_0 = x_{-1} = \dots = x_{-k} = 0$), nonzero $w_t \in L_2[0, \infty)$, and $\hat{V}_{i+1}(x_{i-k}) = 0$, and so (2.60) reduces to $\frac{1}{\gamma}\sum_{t=0}^{\infty}z_t^T z_t - \gamma\sum_{t=0}^{\infty}w_t^T w_t < 0$, which implies that (2.4) is fulfilled.

Next, we show that the closed-loop system (2.5) is globally asymptotically stable when the disturbance $w_t = 0$. Assuming the zero disturbance in (2.58), we have $\Delta_k V(x_t) + \frac{1}{\gamma}\sum_{i=0}^{k}z_{t+i}^T z_{t+i} < 0$. This implies $\Delta_k V(x_t) < 0$. According to Lemma 2.2, the closed-loop system (2.5) is globally asymptotically stable. This completes the proof. $\qquad\square$

Remark 2.2. Theorem 2.1 with $k > 1$ gives more chance to synthesize a controller with H_∞ performance for such a T-S fuzzy systems with standard Lyapunov theory. This means that this methodology is less conservative. However, with increasing k, the number of LMIs and variables will increase.

Remark 2.3. In the literature, the form of LF is usually studied to reduce the conservatism such as common quadratic, piecewise, and fuzzy LF, in which they have shown that a fuzzy Lyapunov form is the least conservative form. In the synthesis method in Theorem 2.2, it is the form of a fuzzy LF. In addition, instead of standard Lyapunov theory, in which LF is decreased in each sample, in this study this condition is also relaxed.

2.4 SIMULATION RESULT

Consider the two-rule model of the uncertain T-S fuzzy system in the form of (2.1), where $A_1 = \begin{bmatrix} -1.9 & -3.5 \\ 2 & 1 \end{bmatrix}$, $A_2 = \begin{bmatrix} -0.08 & 0.8 \\ 0.58 & 0.69 \end{bmatrix}$, $B_{1,1} = \begin{bmatrix} 0.26 \\ 0.92 \end{bmatrix}$, $B_{1,2} = \begin{bmatrix} -0.9 \\ 0.21 \end{bmatrix}$, $D_1 = D_2 = \begin{bmatrix} 0 \\ 0.1 \end{bmatrix}$, $C_1 = C_2 = \begin{bmatrix} 0 & 1 \\ 1 & 0 \end{bmatrix}$, $B_{2,1} = B_{2,2} = \begin{bmatrix} 0 \\ 0 \end{bmatrix}$, $H = 0.1I_2$, $E_1^a = E_2^a = 0.05I_2$, $E_1^b = \begin{bmatrix} 0.06 \\ 0 \end{bmatrix}$, $E_2^b = \begin{bmatrix} -0.05 \\ 0 \end{bmatrix}$.

The parameters ρ_1, ρ_2, and ρ_3 are all selected to be 0.2. The H_∞ robust state feedback controller is designed using Theorem 2.2. The following three cases are considered.

Case 1: Quadratic function $\hat{V}_i(x_t) = x_t^T G^{-T} P_i G^{-1} x_t$.
Case 2: Quadratic fuzzy function $\hat{V}_i(x_t) = x_t^T G^{-T} P_i(v_t) G^{-1} x_t$.
Case 3: Nonquadratic function $\hat{V}_i(x_t) = x_t^T G^{-T}(v_t) P_i(v_t) G^{-1}(v_t) x_t$.

The optimal H_∞ performance, γ_{min}, is computed for each of the aforementioned cases with $k = 1, 2$, and 3. It is worth noting that $k = 1$ corresponds to the standard Lyapunov theory. Table 2.1 shows the optimal γ_{min} for all three cases with different k. It can be observed that γ_{min} decreases as we consider a more general form of $\hat{V}_i(x_t)$. As expected, it is noticed that γ_{min} decreases with increasing the value of k in all three cases. Indeed, gaining smaller γ_{min} implies that the result is less conservative, which clearly verifies our theoretical results. Fig. 2.2 shows $\hat{V}_1(x_t)$, $\hat{V}_2(x_t)$, and $\hat{V}_3(x_t)$ for Case 3. From the figure we observe that $\hat{V}_3(x_t) > 0$, but $\hat{V}_1(x_t)$ and $\hat{V}_2(x_t)$ are not positive-definite.

The designed controller in Case 3 for $k = 3$ is associated with the system and efficiently stabilizes the system states. For simulations, $v_1(t) = x_1(t)$ is

TABLE 2.1 γ_{min} for each case.

	$k = 1$	$k = 2$	$k = 3$
Case 1	2.47	2.19	2
Case 2	2.12	1.92	1.87
Case 3	1.94	1.76	1.7

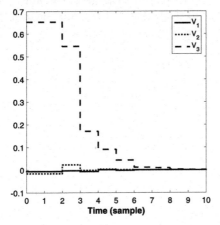

FIGURE 2.2 $\hat{V}_1(x_t)$, $\hat{V}_2(x_t)$, and $\hat{V}_3(x_t)$ in Case 3 with $k = 3$.

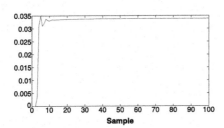

FIGURE 2.3   The ratio of the controlled output energy to the disturbance (γ_{min}) for $k = 3$.

the premise variable, and $\mu_1(\upsilon_1(t)) = 0.5 + 0.5\cos(\upsilon_1(t))$ and $\mu_2(\upsilon_1(t)) = 1 - \mu_1(\upsilon_1(t))$ with sampling time 0.01. The obtained H_∞ performance is 0.0339, as shown in Fig. 2.3, which is the ratio of the controlled output energy z_t to the disturbance energy

$$w_t = \begin{cases} 0.1, & t \text{ even,} \\ -0.1, & t \text{ odd.} \end{cases}$$

This implies that the L_2 gain from the disturbance to the regulated output is not greater than $\gamma_{min} = 0.0339$ and is much less than prescribed value of $\gamma_{min} = 1.7$.

2.5 CONCLUSIONS

In this chapter, based on the NLF approach, we have given sufficient conditions for the existence of a robust H_∞ state feedback controller for uncertain Takagi–Sugeno (T-S) fuzzy systems in terms of linear matrix inequalities (LMI). To the best knowledge of the authors, this is the first report that gives an analytical design of H_∞ robust controller for uncertain discrete-time T-S fuzzy systems using an NLF. The design technique has been shown to be less conservative than the existing k-sample variations of LF via numerical examples. Further research will be aimed at reducing the number of LMIs.

References

[1] T. Takagi, M. Sugeno, Fuzzy identification of systems and its applications to modeling and control, IEEE Transactions on Systems, Man and Cybernetics SMC-15 (1) (Jan 1985) 116–132.

[2] E.F. Stephen Boyd, Laurent El Ghaoui, V. Balakrishnan, Linear Matrix Inequalities in System and Control Theory, Society for Industrial and Applied Mathematics (SIAM), 1994.

[3] G. Feng, A survey on analysis and design of model-based fuzzy control systems, IEEE Transactions on Fuzzy Systems 14 (5) (Oct 2006) 676–697.

[4] A. Sala, T.M. Guerra, R. Babuska, Perspectives of fuzzy systems and control, Fuzzy Sets and Systems 156 (3) (2005) 432–444.

[5] H. Li, Y. Gao, P. Shi, X. Zhao, Output-feedback control for T-S fuzzy delta operator systems with time-varying delays via an input–output approach, IEEE Transactions on Fuzzy Systems 23 (4) (Aug 2015) 1100–1112.

[6] H. Li, Y. Pan, Z. Yu, X. Zhao, X. Yang, Fuzzy output-feedback control for non-linear systems with input time-varying delay, IET Control Theory & Applications 8 (9) (June 2014) 738–745.

[7] G. Feng, Stability analysis of discrete-time fuzzy dynamic systems based on piecewise Lyapunov functions, IEEE Transactions on Fuzzy Systems 12 (1) (Feb 2004) 22–28.

[8] K. Tanaka, H. Yoshida, H. Ohtake, H. Wang, A sum-of-squares approach to modeling and control of nonlinear dynamical systems with polynomial fuzzy systems, IEEE Transactions on Fuzzy Systems 17 (4) (Aug 2009) 911–922.

[9] H.-N. Wu, K.-Y. Cai, H_2 guaranteed cost fuzzy control design for discrete-time nonlinear systems with parameter uncertainty, Automatica 42 (7) (July 2006) 1183–1188.

[10] H. Li, C. Wu, L. Wu, H.K. Lam, Y. Gao, Filtering of interval type-2 fuzzy systems with intermittent measurements, IEEE Transactions on Cybernetics 46 (3) (March 2016) 668–678.

[11] H. Li, J. Wang, H.K. Lam, Q. Zhou, H. Du, Adaptive sliding mode control for interval type-2 fuzzy systems, IEEE Transactions on Systems, Man, and Cybernetics: Systems 46 (12) (Dec 2016) 1654–1663.

[12] A. Ahmadi, P. Parrilo, Non-monotonic Lyapunov functions for stability of discrete time nonlinear and switched systems, in: 47th IEEE Conference on Decision and Control, 2008 (CDC 2008), Dec 2008, pp. 614–621.

[13] S.F. Derakhshan, A. Fatehi, Non-monotonic Lyapunov functions for stability analysis and stabilization of discrete time Takagi–Sugeno fuzzy systems, International Journal of Innovative Computing, Information & Control (2014) 1567–1586.

[14] S. Derakhshan, A. Fatehi, M. Sharabiany, Nonmonotonic observer-based fuzzy controller designs for discrete time T-S fuzzy systems via LMI, IEEE Transactions on Cybernetics (99) (April 2014).

[15] S.F. Derakhshan, A. Fatehi, Non-monotonic robust H_2 fuzzy observer-based control for discrete time nonlinear systems with parametric uncertainties, International Journal of Systems Science 46 (12) (2015) 2134–2149.

[16] T. Guerra, A. Kruszewski, M. Bernal, Control law proposition for the stabilization of discrete Takagi–Sugeno models, IEEE Transactions on Fuzzy Systems 17 (3) (June 2009) 724–731.

[17] A. Kruszewski, R. Wang, T. Guerra, Nonquadratic stabilization conditions for a class of uncertain nonlinear discrete time T-S fuzzy models: a new approach, IEEE Transactions on Automatic Control 53 (2) (March 2008) 606–611.

[18] Y.-J. Chen, M. Tanaka, K. Inoue, H. Ohtake, K. Tanaka, T. Guerra, A. Kruszewski, H. Wang, A nonmonotonically decreasing relaxation approach of Lyapunov functions to guaranteed cost control for discrete fuzzy systems, IET Control Theory & Applications 8 (16) (2014) 1716–1722.

[19] E. Kim, H. Lee, New approaches to relaxed quadratic stability condition of fuzzy control systems, IEEE Transactions on Fuzzy Systems 8 (5) (Oct 2000) 523–534.

[20] W.-J. Wang, Y.-J. Chen, C.-H. Sun, Relaxed stabilization criteria for discrete-time T-S fuzzy control systems based on a switching fuzzy model and piecewise Lyapunov function, IEEE Transactions on Systems, Man and Cybernetics. Part B. Cybernetics 37 (3) (June 2007) 551–559.

[21] D.J. Choi, P. Park, H_∞ state-feedback controller design for discrete-time fuzzy systems using fuzzy weighting-dependent Lyapunov functions, IEEE Transactions on Fuzzy Systems 11 (2) (Apr 2003) 271–278.

[22] M.C.D. Oliveora, J.C. Geromel, J. Berussou, Extended H_2 and H norm characterizations and controller parametrizations for discrete-time systems, International Journal of Control 75 (9) (2002) 666–679.

[23] H. Tuan, P. Apkarian, T. Narikiyo, Y. Yamamoto, Parameterized linear matrix inequality techniques in fuzzy control system design, IEEE Transactions on Fuzzy Systems 9 (2) (2001) 324–332.

Robust H_∞ Filtering of T-S Fuzzy Systems: A Nonmonotonic Approach

3.1 INTRODUCTION

It is well known that most of practical systems are inherently nonlinear, and one of the widely used techniques to approximate nonlinear systems is the Takagi–Sugeno (T-S) fuzzy model [1]. The T-S fuzzy model is a combination of linear models that are connected together by nonlinear scalar membership functions. Such a minimum complexity feature of the T-S fuzzy model has attracted large attention for H_∞ filtering problem. A comprehensive review of various types of filters, which have been designed for T-S fuzzy systems, can be found in [2] and references therein.

Generally speaking, the stability analysis of the T-S model and subsequently filter design are often carried out through the standard Lyapunov method. This inherently leads to some inequalities in which the nonlinear membership functions are excluded from final conditions. As a result, the obtained conditions, which are often constructed in the form of linear matrix inequalities (LMIs), become only sufficient (conservative). In the literature, two main approaches have been presented to mitigate the demerits of this conservatism. The first approach focuses on judiciously choosing a suitable Lyapunov function (LF) to reduce the conservatism, e.g., piecewise [3], fuzzy [4], nonquadratic Lyapunov functions [5], and combinations of these. The second approach, which is the main focus of this study, reduces the conservatism by relaxing the monotonic decrease of LFs in discrete-time systems. This approach is alternately referred to as nonmonotonic approach [6–8]. In other words, LFs in this approach are not required anymore to decrease in each successive step (i.e., $V_{t+1} < V_t$). Instead, it allows small local growth, but it ultimately converges to zero [9]. Recently, some works have been made using the nonmonotonic approach.

43

In [8], the nonmonotonic decrease of a Lyapunov function is based on only 2-sample variations (i.e., $V_{t+2} < V_t$). This nonmonotonic approach has been shown to be less conservative than the common quadratic LF and even piecewise Lyapunov function. A more general nonmonotonic Lyapunov function (NLF) with k-sample variations (i.e., $V_{t+k} < V_t$) has been introduced in [6]. In [6], a robust state feedback controller for uncertain discrete-time T-S fuzzy systems has been designed. Moreover, the same approach is considered for guaranteed cost control design in [7].

Robust H_∞ filtering turns out to be more difficult to solve when compared to the aforecited control problems using NLFs. Thus, it remains challenging and needs to be solved. This chapter extends k-sample variation [6, 8] to the new NLF approach. This new NLF approach introduces more decision variables and inequalities to further reduce the conservatism. Then the design of a robust H_∞ filter, which is the main objective of this chapter, is investigated using the extended NLF approach. Finally, sufficient conditions for the existence of a robust H_∞ filter are derived in terms of LMIs.

The rest of the chapter is organized as follows. Section 3.2 briefly describes the system and problem formulation. Main results are presented in Section 3.3. The effectiveness of the filter design is demonstrated in Section 3.4. Conclusions are drawn in Section 3.5.

Notation: Throughout the chapter, $(\cdot)^T$ and $\bullet$ denote, respectively, the transpose and the symmetric block in a symmetric matrix. The identity matrix of size n is denoted by I_n, and 0_{nm} represents the zero matrix of size $n \times m$; $x_t \triangleq x(t)$ and $\mathbb{K}_{r_1, r_L} \triangleq \{r_1, \ldots, r_L\}$, where $t, r_1, \ldots, r_L \in \mathbb{N}$; $X(v_t) \triangleq \sum_{i=1}^{r} \mu_i(v_t) X_i$ with $\mu_i(v_t) \geq 0$ and $\sum_{i=1}^{r} \mu_i(v_t) = 1$.

3.2 SYSTEM DESCRIPTION AND PROBLEM FORMULATION

This chapter considers the following uncertain discrete-time T-S fuzzy system:

Plant Rule i: IF $v_{1,t}$ is M_1^i AND $\cdots$ AND $v_{\sigma,t}$ is M_σ^i, THEN

$$
\begin{aligned}
x_{t+1} &= (A_i + \Delta A_i) x_t + D_{x,i} w_t, \\
y_t &= C_{y,i} x_t + D_{y,i} w_t, \\
z_t &= C_{z,i} x_t + B_{z,i} u_t,
\end{aligned}
\tag{3.1}
$$

where $x_t \in \mathbb{R}^n$, $z_t \in \mathbb{R}^p$, $y_t \in \mathbb{R}^{p_y}$, and $w_t \in \mathbb{R}^d$ are the state, control, regulated output, measured output signals, and disturbance, respectively, $i \in \mathbb{K}_{1,r}$ denotes the ith fuzzy inference rule, r is the number of rules, $v_{\tau,t}$ are the premise variables ($v_t = [v_{1,t}, \cdots, v_{\sigma,t}]$), σ is the number of premise variables, and $M_1^i, \ldots, M_\sigma^i$ are membership functions. The matrices $A_i, \Delta A_i, D_{x,i}, D_{y,i}, C_{y,i}, C_{z,i}$, and $B_{z,i}$ are of appropriate dimensions. By using a center-average defuzzifier, product inference, and singleton fuzzifier, the uncertain T-S fuzzy system in (3.1) can be expressed as follows:

$$
\begin{aligned}
x_{t+1} &= (A(v_t) + \Delta A(v_t)) x_t + D_x(v_t) w_t, \\
y_t &= C_y(v_t) x_t + D_y(v_t) w_t, \\
z_t &= C_z(v_t) x_t + B_z(v_t) u_t,
\end{aligned}
\tag{3.2}
$$

where $\mu_i(v_t) = \frac{\prod_{\tau=1}^{\sigma} M_\tau^i(v_{\tau,t})}{\sum_{\ell=1}^{r} \prod_{\tau=1}^{\sigma} M_\tau^\ell(v_{\tau,t})}$ with $\mu_i(v_t) \geq 0$ and $\sum_{i=1}^{r} \mu_i(v_t) = 1$. The system uncertainties are assumed to be $\Delta A(v_t) = H F(x_t, t) E^a(v_t)$, where H are $E^a(v_t)$ are known matrix functions that characterize the structure of the uncertainties, and $F(x_t, t)$ is an unknown nonlinear time-varying matrix function satisfying $F(x_t, t)^T F(x_t, t) \leq I$. The full-order PDC filter in the form of

$$
\begin{aligned}
\widehat{x}_{t+1} &= \sum_{j=1}^{r} \left\{ \widehat{A}_j \widehat{x}_t + \widehat{B}_j y_t \right\} \\
&= \widehat{A}(v_t) \widehat{x}_t + \widehat{B}(v_t) y_t, \\
\widehat{z}_t &= \sum_{j=1}^{r} \mu_j(v_t) \left\{ \widehat{C}_j \widehat{x}_t + \widehat{D}_j y_t \right\} \\
&= \widehat{C}(v_t) \widehat{x}_t + \widehat{D}(v_t) y_t
\end{aligned}
\tag{3.3}
$$

is developed, which guarantees that the output of the filtering $\widehat{z}_t$ tracks the output of the system z_t, $\widehat{x}_t \in \mathbb{R}^n$ is the filter state, and $\widehat{A}_j, \widehat{B}_j, \widehat{C}_j$, and $\widehat{D}_j$ are filter matrices of appropriate dimensions to be determined. Augmenting system (3.2) to include the filter states in (3.3), we obtain the following filtering error system:

$$
\begin{aligned}
\check{x}_{t+1} &= \check{A}(v_t) \check{x}_t + \check{B}(v_t) w_t, \\
e_t &= \check{C}(v_t) \check{x}_t + \check{D}(v_t) w_t,
\end{aligned}
\tag{3.4}
$$

where $e_t = \widehat{z}_t - z_t$ is the filtering error, $\check{x}_t^T = \begin{bmatrix} x_t^T & \widehat{x}_t^T \end{bmatrix}$ is the augmented state vector, and

$$
\check{A}(v_t) = \begin{bmatrix} A(v_t) + \Delta A(v_t) & 0 \\ \widehat{B}(v_t) C_y(v_t) & \widehat{A}(v_t) \end{bmatrix},
\tag{3.5}
$$

$$
\check{B}(v_t) = \begin{bmatrix} D_x(v_t) \\ \widehat{B}(v_t) D_y(v_t) \end{bmatrix},
\tag{3.6}
$$

$$\check{C}(v_t) = \begin{bmatrix} -C_z(v_t) + \widehat{D}(v_t)C_y(v_t) & \widehat{C}(v_t) \end{bmatrix}, \tag{3.7}$$

$$\check{D}(v_t) = \widehat{D}(v_t)D_y(v_t). \tag{3.8}$$

Robust H_∞ Filtering Problem Formulation: The aim of the robust H_∞ filtering problem is to estimate z_t such that the filtering error system (3.4) with $w_t = 0$ is globally asymptotically stable and the filtering error system (3.4) guarantees

$$\sum_{t=0}^{\infty} e_t^T e_t < \gamma \sum_{t=0}^{\infty} \left(w_t^T w_t \right) \tag{3.9}$$

under zero initial conditions for all nonzero $w_t \in L_2[0, \infty)$ and a given positive constant scalar γ.

Before proceeding to the next section, let recall the following lemma.

Lemma 3.1. *[14] For any positive scalar ρ_1, matrices Υ_0 $\tilde{H}$, $\tilde{E}_t$ of appropriate dimensions, and nonlinear time-varying matrix function $\tilde{F}_t$ satisfying $\tilde{F}_t^T \tilde{F}_t \leq I$, we have*

$$\Upsilon_0 + \tilde{H}\tilde{F}_t\tilde{E}_t + \left(\tilde{H}\tilde{F}_t\tilde{E}_t \right)^T \leq \Upsilon_0 + \rho_1 \tilde{H}\tilde{H}^T + \rho_1^{-1}\tilde{E}_t^T \tilde{E}_t. \tag{3.10}$$

3.3 MAIN RESULTS

The following theorem provides sufficient conditions for the existence of a robust H_∞ filtering (3.3) using an NLF.

Theorem 3.1. *The filtering error system (3.4) is globally asymptotically stable with H_∞ norm bound γ if there exist positive scalar parameters $\hat{\rho}_g$, a set of symmetric matrices $\bar{\mathbf{P}}_{g,11}^{ih}$, $\bar{\mathbf{P}}_{g,22}^{ih}$, $\check{\Phi}_{L,11}^{i_0 j_0 \ldots i_L j_L}$, $\check{\Phi}_{L,22}^{i_0 j_0 \ldots i_L j_L}$ and matrices X, Y, Z, $\check{\Phi}_{L,21}^{i_0 j_0 \ldots i_L j_L}$, $\bar{\mathbf{P}}_{g,21}^{ih}$ for $h \in \mathbb{K}_{0,k-1}$, $i_0 \ldots i_k j_0 \ldots j_{k-1} \in \mathbb{K}_{1,r}$, $L \in \mathbb{K}_{1,k-1}$, and $g \in \mathbb{K}_{1,k}$ such that*

$$\sum_{g=j}^{k} \bar{\mathbf{P}}_{g,i_h} > 0 \qquad \forall i_h \in \mathbb{K}_{1,r} \text{ and } j \in \mathbb{K}_{1,k}, \tag{3.11}$$

$$\begin{cases} \Psi_{g,i_{k-g}i_{k-g}} < 0, \\ \frac{1}{r-1}\Psi_{g,i_{k-g}i_{k-g}} + \frac{1}{2}\left(\Psi_{g,i_{k-g}j_{k-g}} + \Psi_{g,j_{k-g}i_{k-g}} \right) < 0 \quad \forall i_{k-g} \neq j_{k-g}, \end{cases} \tag{3.12}$$

$$\bar{\mathbf{P}}_{g,i_h} = \begin{bmatrix} \bar{\mathbf{P}}_{g,i_h}^{11} & \bullet \\ \bar{\mathbf{P}}_{g,i_h}^{21} & \bar{\mathbf{P}}_{g,i_h}^{11} \end{bmatrix}, \tag{3.13}$$

$$
\Psi_{g,i_{k-g}j_{k-g}} = \begin{bmatrix}
\Psi_g^{11} & \bullet & \bullet \\
\Psi_g^{21} & \Psi_g^{22} & \bullet \\
0_{dn} & 0_{dn} & -\gamma I_d \\
-C_{z,i_{k-g}}X + \mathbf{C}_{jk-g} & -C_{z,i_{k-g}} + \mathbf{D}_{jk-g}C_{y,i_{k-g}} & \mathbf{D}_{jk-g}D_{y,i_{k-g}} \\
A_{i_{k-g}}X & A_{i_{k-g}} & D_{x,i_{k-g}} \\
\mathbf{A}_{jk-g} & YA_{i_{k-g}} + \mathbf{B}_{jk-g}C_{y,i_{k-g}} & YD_{x,i_{k-g}} + \mathbf{B}_{jk-g}D_{y,i_{k-g}} \\
E_{i_{k-g}}^a X & E_{i_{k-g}}^a & 0_{nd} \\
0_{nn} & 0_{nn} & 0_{nd} \\
0_{nn} & 0_{nn} & 0_{nd} \\
0_{nn} & 0_{nn} & 0_{nd}
\end{bmatrix}
$$

$$
\begin{bmatrix}
\bullet & \bullet & \bullet & \bullet & \bullet & \bullet & \bullet \\
\bullet & \bullet & \bullet & \bullet & \bullet & \bullet & \bullet \\
\bullet & \bullet & \bullet & \bullet & \bullet & \bullet & \bullet \\
-\gamma I_p & \bullet & \bullet & \bullet & \bullet & \bullet & \bullet \\
0_{np} & \Psi_g^{55} & \bullet & \bullet & \bullet & \bullet & \bullet \\
0_{np} & \Psi_g^{65} & \Psi_g^{66} & \bullet & \bullet & \bullet & \bullet \\
0_{np} & 0_{nn} & 0_{nn} & -\hat{\rho}_g I_n & \bullet & \bullet & \bullet \\
0_{np} & 0_{nn} & 0_{nn} & 0_{nn} & -\hat{\rho}_g I_n & \bullet & \bullet \\
0_{np} & H^T & H^T Y^T & 0_{nn} & 0_{nn} & -\hat{\rho}_g^{-1} I_n & \bullet \\
0_{np} & 0_{nn} & 0_{nn} & 0_{nn} & 0_{nn} & 0_{nn} & -\hat{\rho}_g^{-1} I_n
\end{bmatrix}, \quad (3.14)
$$

$$\Psi_1^{11} = \check{\Phi}_{k-1,11}^{i_0 j_0 \ldots i_{k-1} j_{k-1}} - \bar{\mathbf{P}}_{k-1,11}^{i_{k-1}}, \quad \Psi_1^{21} = \check{\Phi}_{k-1,21}^{i_0 j_0 \ldots i_{k-1} j_{k-1}} - \bar{\mathbf{P}}_{k-1,21}^{i_{k-1}},$$

$$\Psi_1^{22} = \check{\Phi}_{k-1,22}^{i_0 j_0 \ldots i_{k-1} j_{k-1}} - \bar{\mathbf{P}}_{k-1,22}^{i_{k-1}},$$

$$\Psi_1^{55} = -X - X^T + \bar{\mathbf{P}}_{k,11}^{i_k}, \quad \Psi_1^{65} = -I_n - Z + \bar{\mathbf{P}}_{k,21}^{i_k}, \quad \Psi_1^{66} = -Y - Y^T + \bar{\mathbf{P}}_{k,22}^{i_k},$$

$$\Psi_s^{11} = \check{\Phi}_{k-s,11}^{i_0 j_0 \ldots i_{k-s} j_{k-s}} - \bar{\mathbf{P}}_{k-s,11}^{i_{k-s}}, \quad \Psi_s^{21} = \check{\Phi}_{k-s,21}^{i_0 j_0 \ldots i_{k-s} j_{k-s}} - \bar{\mathbf{P}}_{k-s,21}^{i_{k-s}}, \quad \Psi_s^{22}$$

$$= \check{\Phi}_{k-s,22}^{i_0 j_0 \ldots i_{k-s} j_{k-s}} - \bar{\mathbf{P}}_{k-s,22}^{i_{k-s}},$$

$$\Psi_s^{55} = -X - X^T + \check{\Phi}_{k-s+1,11}^{i_0 j_0 \ldots i_{k-s+1} j_{k-s+1}}, \quad \Psi_s^{65} = -I_n - Z + \check{\Phi}_{k-s+1,21}^{i_0 j_0 \ldots i_{k-s+1} j_{k-s+1}},$$

$$\Psi_s^{66} = -Y - Y^T + \check{\Phi}_{k-s+1,22}^{i_0 j_0 \ldots i_{k-s+1} j_{k-s+1}}, \quad s \in \mathbb{K}_{2,k-1},$$

$$\Psi_k^{11} = -\sum_{i=1}^{k} \bar{\mathbf{P}}_{i,11}^{i_0}, \quad \Psi_k^{21} = -\sum_{i=1}^{k} \bar{\mathbf{P}}_{i,21}^{i_0}, \quad \Psi_k^{22} = -\sum_{i=1}^{k} \bar{\mathbf{P}}_{i,22}^{i_0},$$

$$\Psi_k^{55} = -X - X^T + \check{\Phi}_{1,11}^{i_0 j_0 i_1 j_1}, \quad \Psi_k^{65} = -I_n - Z + \check{\Phi}_{1,21}^{i_0 j_0 i_1 j_1},$$

$$\Psi_k^{66} = -Y - Y^T + \check{\Phi}_{1,22}^{i_0 j_0 i_1 j_1}.$$

The filter matrices $\widehat{A}_{i_h} \in \mathbb{R}^{n \times n}$, $\widehat{B}_{i_h} \in \mathbb{R}^{n \times p_y}$, $\hat{C}_{i_h} \in \mathbb{R}^{p \times n}$, *and* $\widehat{D}_{i_h} \in \mathbb{R}^{p \times p_y}$ *in (3.3) are given as follows:*

$$\begin{bmatrix} \widehat{A}_{ih} & \widehat{B}_{ih} \\ \widehat{C}_{ih} & \widehat{D}_{ih} \end{bmatrix} = \begin{bmatrix} M^{-1} & 0 \\ 0 & I \end{bmatrix} \begin{bmatrix} \mathbf{A}_{ih} - YA_{ih}X & \mathbf{B}_{ih} \\ \mathbf{C}_{ih} & \mathbf{D}_{ih} \end{bmatrix}$$

$$\times \begin{bmatrix} N^{-1} & 0 \\ -C_{y,ih}XN^{-1} & I \end{bmatrix}.$$

Note that the two nonsingular constant matrices $N \in \mathbb{R}^{n \times n}$ and $M \in \mathbb{R}^{n \times n}$ can always be obtained such that $MN = Z - YX$.

Proof. Consider the function

$$\check{V}_g(\check{x}_t) \triangleq \check{x}_t^T \check{P}_g(v_t)\check{x}_t \quad \forall\, g \in \mathbb{K}_{1,k}, \tag{3.15}$$

where

$$\check{P}_g(v_t) = G^{-T}P_g(v_t)G^{-1}. \tag{3.16}$$

Let us define

$$\Delta_k \check{V}(\check{x}_t) \triangleq \sum_{g=1}^{k} \left(\check{x}_{t+g}^T \check{P}_g(v_{t+g})\check{x}_{t+g} - \check{x}_t^T \check{P}_g(v_t)\check{x}_t \right), \tag{3.17}$$

$$\Pi_L \triangleq \sum_{i_L=1}^{r} \sum_{j_L=1}^{r} \cdots \sum_{i_0=1}^{r} \sum_{j_0=1}^{r} \mu_{i_0}(v_t)\mu_{j_0}(v_t)\ldots\mu_{i_L}(v_{t+L})\mu_{j_L}(v_{t+L})\Phi_{L,i_0 j_0..i_L j_L},$$

$$\tag{3.18}$$

$$\check{\Pi}_L \triangleq G^{-T}\Pi_L G^{-1}, \tag{3.19}$$

which in the following steps help us to show that the theorem conditions satisfy the H_∞ performance and stability of the error filtering system.

Step 1: Adding and subtracting $\check{x}_{t+k-1}^T \check{\Pi}_{k-1}\check{x}_{t+k-1} + \frac{1}{\gamma}e_{t+k-1}^T e_{t+k-1} - \gamma w_{t+k-1}^T w_{t+k-1}$ to and from $\Delta_k \check{V}(\check{x}_t)$ defined in (3.17) and rearranging its terms give

$$\begin{aligned} \Delta_k \check{V}(\check{x}_t) &= \check{x}_{t+k}^T \check{P}_k(v_{t+k})\check{x}_{t+k} + \check{x}_{t+k-1}^T \check{P}_{k-1}(v_{t+k-1})\check{x}_{t+k-1} \\ &\quad -\check{x}_{t+k-1}^T \check{\Pi}_{k-1}\check{x}_{t+k-1} + \frac{1}{\gamma}e_{t+k-1}^T e_{t+k-1} - \gamma w_{t+k-1}^T w_{t+k-1} \\ &\quad +\check{x}_{t+k-1}^T \check{\Pi}_{k-1}\check{x}_{t+k-1} + \Delta_{k-2}\check{V}(\check{x}_t) - \check{x}_t^T \left(\sum_{g=k-1}^{k} \check{P}_g(v_t) \right) \check{x}_t \\ &\quad -\frac{1}{\gamma}e_{t+k-1}^T e_{t+k-1} + \gamma w_{t+k-1}^T w_{t+k-1}. \end{aligned} \tag{3.20}$$

Note that from (3.15) and (3.17) we have $\Delta_k \check{V}(\check{x}_t) = \Delta_{k-2}\check{V}(\check{x}_t) + \check{x}_{t+k}^T \check{P}_k(v_{t+k})\check{x}_{t+k} + \check{x}_{t+k-1}^T \check{P}_{k-1}(v_{t+k-1})\check{x}_{t+k-1} - \check{x}_t^T \left(\sum_{g=k-1}^{k} \check{P}_g(v_t) \right) \check{x}_t$.

Define $\zeta_t^T \triangleq \begin{bmatrix} \check{x}_t^T & w_t^T \end{bmatrix}$. Using the system equation (3.4), we have $\check{x}_{t+k} = \check{A}(v_{t+k-1})\check{x}_{t+k-1} + \check{B}(v_{t+k-1})w_{t+k-1}$ and $e_{t+k-1} = \check{C}(v_{t+k-1})\check{x}_{t+k-1} + \check{D}(v_{t+k-1})w_{t+k-1}$. Substituting these into (3.20) gives

$$
\begin{aligned}
\Delta_k \check{V}(\check{x}_t) &= \zeta_{t+k-1}^T \check{\chi}_1 \zeta_{t+k-1} + \check{x}_{t+k-1}^T \check{\Pi}_{k-1} \check{x}_{t+k-1} + \Delta_{k-2}\check{V}(\check{x}_t) \\
&\quad - \frac{1}{\gamma} e_{t+k-1}^T e_{t+k-1} + \gamma w_{t+k-1}^T w_{t+k-1} \\
&\quad - \check{x}_t^T \left(\sum_{g=k-1}^k \check{P}_g(v_t) \right) \check{x}_t,
\end{aligned}
\tag{3.21}
$$

where

$$
\check{\chi}_1 = \begin{bmatrix} \check{\chi}_1^{11} & \check{\chi}_1^{12} \\ \bullet & \check{\chi}_1^{22} \end{bmatrix},
\tag{3.22}
$$

$$
\begin{aligned}
\check{\chi}_1^{11} &= \check{A}^T(v_{t+k-1})\check{P}_k(v_{t+k})\check{A}(v_{t+k-1}) - \check{\Pi}_{k-1} + \check{P}_{k-1}(v_{t+k-1}) \\
&\quad + \frac{1}{\gamma}\check{C}^T(v_{t+k-1})\check{C}(v_{t+k-1}), \\
\check{\chi}_1^{12} &= \check{A}^T(v_{t+k-1})\check{P}_k(v_{t+k})\check{B}(v_{t+k-1}) + \frac{1}{\gamma}\check{C}^T(v_{t+k-1})\check{D}(v_{t+k-1}), \\
\check{\chi}_1^{22} &= \check{B}^T(v_{t+k-1})\check{P}_k(v_{t+k})\check{B}(v_{t+k-1}) - \gamma I_d + \frac{1}{\gamma}\check{D}^T(v_{t+k-1})\check{D}(v_{t+k-1}).
\end{aligned}
\tag{3.23}
$$

In the rest of this step, we show that if (3.12) for $g = 1$ holds, then $\check{\chi}_1$ is less than zero. Pre- and postmultiplying (3.22) by $diag\{G^T, I\}$ and its transpose, respectively, gives

$$
\Xi_1 = \begin{bmatrix} \Xi_1^{11} & \Xi_1^{12} \\ \bullet & \check{\chi}_1^{22} \end{bmatrix},
\tag{3.24}
$$

where

$$
\begin{aligned}
\Xi_1^{11} &= G^T \check{A}^T(v_{t+k-1})\check{P}_k(v_{t+k})\check{A}(v_{t+k-1})G - \Pi_{k-1} + P_{k-1}(v_{t+k-1}) \\
&\quad + \frac{1}{\gamma}G^T \check{C}^T(v_{t+k-1})\check{C}(v_{t+k-1})G, \\
\Xi_1^{12} &= G^T \check{A}^T(v_{t+k-1})\check{P}_k(v_{t+k})\check{B}(v_{t+k-1}) + \frac{1}{\gamma}G^T \check{C}^T(v_{t+k-1})\check{D}(v_{t+k-1}).
\end{aligned}
\tag{3.25}
$$

So, if $\varXi_1 < 0$, then $\check\chi_1 < 0$. Using Schur's complement, $\varXi_1 < 0$ holds if

$$
\begin{bmatrix}
-\Pi_{k-1} + P_{k-1}(v_{t+k-1}) & \bullet & \bullet & \bullet \\
0 & -\gamma I & \bullet & \bullet \\
\check{C}(v_{t+k-1})G & \check{D}(v_{t+k-1}) & -\gamma I & \bullet \\
\check{A}(v_{t+k-1})G & \check{B}(v_{t+k-1}) & 0 & -\check{P}_{t+k}^{-1}
\end{bmatrix} < 0. \qquad (3.26)
$$

According to Lemma 3.1, $-\check{P}_{t+k}^{-1} \leq -G^T - G + P_k(v_{t+k})$, and therefore (3.26) holds if

$$
\Upsilon_1 =
\begin{bmatrix}
-\Pi_{k-1} + P_{k-1}(v_{t+k-1}) & \bullet & \bullet & \bullet \\
0 & -\gamma I & \bullet & \bullet \\
\check{C}(v_{t+k-1})G & \check{D}(v_{t+k-1}) & -\gamma I & \bullet \\
\check{A}(v_{t+k-1})G & \check{B}(v_{t+k-1}) & 0 & -G^T - G + P_k(v_{t+k})
\end{bmatrix} < 0. \qquad (3.27)
$$

Define $\check{A}(v_t) \triangleq \bar{A}(v_t) + \bar{H}\bar{F}_t\bar{E}_t^a$,

$$
\tilde{H} \triangleq
\begin{bmatrix}
0 & 0 & 0 & 0 \\
0 & 0 & 0 & 0 \\
0 & 0 & 0 & 0 \\
\bar{H} & 0 & 0 & 0
\end{bmatrix}, \qquad (3.28)
$$

$$
\tilde{F} \triangleq
\begin{bmatrix}
\bar{F} & 0 & 0 & 0 \\
0 & 0 & 0 & 0 \\
0 & 0 & 0 & 0 \\
0 & 0 & 0 & 0
\end{bmatrix}, \qquad (3.29)
$$

$$
\tilde{E}_t \triangleq
\begin{bmatrix}
\bar{E}_t^a G & 0 & 0 & 0 \\
0 & 0 & 0 & 0 \\
0 & 0 & 0 & 0 \\
0 & 0 & 0 & 0
\end{bmatrix}, \qquad (3.30)
$$

where

$$
\bar{A}(v_t) =
\begin{bmatrix}
A(v_t) & 0 \\
\widehat{B}(v_t)C_y(v_t) & \widehat{A}(v_t, v_t)
\end{bmatrix}, \qquad (3.31)
$$

$$
\bar{H} =
\begin{bmatrix}
H & 0 \\
0 & 0
\end{bmatrix}, \qquad (3.32)
$$

$$
\bar{F}_t =
\begin{bmatrix}
F_t & 0 \\
0 & 0
\end{bmatrix}, \qquad (3.33)
$$

$$\bar{E}_t^a = \begin{bmatrix} E^a(v_t) & 0 \\ 0 & 0 \end{bmatrix}. \tag{3.34}$$

Then Υ_1 can be expressed as

$$\Upsilon_1 = \Upsilon_0 + \tilde{H}\tilde{F}_{t+k-1}\tilde{E}_{t+k-1} + \left(\tilde{H}\tilde{F}_{t+k-1}\tilde{E}_{t+k-1}\right)^T, \tag{3.35}$$

where

$$\Upsilon_0 \triangleq \begin{bmatrix} -\Pi_{k-1} + P_{k-1}(v_{t+k-1}) & \bullet & \bullet & \bullet \\ 0 & -\gamma I & \bullet & \bullet \\ \check{C}(v_{t+k-1})G & \check{D}(v_{t+k-1}) & -\gamma I & \bullet \\ \bar{A}(v_{t+k-1})G & \check{B}(v_{t+k-1}) & 0 & \begin{pmatrix} -G^T - G \\ +P_k(v_{t+k}) \end{pmatrix} \end{bmatrix}. \tag{3.36}$$

Using Lemma 2.3 in Chapter 2, (3.27) holds if

$$\begin{bmatrix} \begin{pmatrix} -\Pi_{k-1} + P_{k-1}(v_{t+k-1}) \\ +\rho_1^{-1}G^T \bar{E}_{t+k-1}^{aT} \bar{E}_{t+k-1}^a G \end{pmatrix} & \bullet & \bullet & \bullet \\ 0 & -\gamma I & \bullet & \bullet \\ \check{C}(v_{t+k-1})G & \check{D}(v_{t+k-1}) & -\gamma I & \bullet \\ \bar{A}(v_{t+k-1})G & \check{B}(v_{t+k-1}) & 0 & -G^T - G + P_k(v_{t+k}) + \rho_1 \tilde{H}\tilde{H}^T \end{bmatrix} < 0. \tag{3.37}$$

Using Schur's complement on (3.37), it is equivalent to

$$\begin{bmatrix} -\Pi_{k-1} + P_{k-1}(v_{t+k-1}) & \bullet & \bullet & \bullet & \bullet & \bullet \\ 0 & -\gamma I & \bullet & \bullet & \bullet & \bullet \\ \check{C}(v_{t+k-1})G & \check{D}(v_{t+k-1}) & -\gamma I & \bullet & \bullet & \bullet \\ \bar{A}(v_{t+k-1})G & \check{B}(v_{t+k-1}) & 0 & -G^T - G + P_k(v_{t+k}) & \bullet & \bullet \\ \bar{E}^a(v_{t+k-1})G & 0 & 0 & 0 & -\rho_1 I & \bullet \\ 0 & 0 & 0 & \tilde{H}^T & 0 & -\rho_1^{-1}I \end{bmatrix} < 0. \tag{3.38}$$

Until here it has been shown that if (3.38) holds, then $\check{\chi}_1$ is less than zero. Let define the matrices G, G^{-1}, S, $P_g(v_{t+g})$, and Π_L partitioned into blocks as

$$G \triangleq \begin{bmatrix} X & T_1 \\ N & T_2 \end{bmatrix}, \quad G^{-1} \triangleq \begin{bmatrix} Y^T & T_3 \\ M^T & T_4 \end{bmatrix}, \quad S \triangleq \begin{bmatrix} I & Y^T \\ 0 & M^T \end{bmatrix}, \tag{3.39}$$

$$P_g(v_{t+g}) \triangleq \begin{bmatrix} P_{g,11}(v_{t+g}) & \bullet \\ P_{g,21}(v_{t+g}) & P_{g,22}(v_{t+g}) \end{bmatrix}, \tag{3.40}$$

$$\Pi_L \triangleq \begin{bmatrix} \Pi_{L,11} & \bullet \\ \Pi_{L,21} & \Pi_{L,22} \end{bmatrix}, \tag{3.41}$$

where T_1, T_2, T_3, and T_4 in these matrices are not important in the sequel [13]. Let define the following nonlinear transformation (change of variables), which helps us in linearizing the matrix inequality in (3.38):

$$S^T P_g(v_{t+g}) S \triangleq \begin{bmatrix} \bar{P}_{g,11}(v_{t+g}) & \bullet \\ \bar{P}_{g,21}(v_{t+g}) & \bar{P}_{g,22}(v_{t+g}) \end{bmatrix}, \tag{3.42}$$

$$S^T \Pi_L S \triangleq \begin{bmatrix} \bar{\Phi}_{L,11} & \bullet \\ \bar{\Phi}_{L,21} & \bar{\Phi}_{L,22} \end{bmatrix}$$

$$= \sum_{i_L=1}^{r} \sum_{j_L=1}^{r} \cdots \sum_{i_0=1}^{r} \sum_{j_0=1}^{r} \mu_{i_0}(v_t) \mu_{j_0}(v_t) \ldots \mu_{i_L}(v_{t+L}) \mu_{j_L}(v_{t+L}) \tag{3.43}$$

$$\times \begin{bmatrix} \check{\Phi}_{L,11}^{i_0 j_0 \ldots i_L j_L} & \bullet \\ \check{\Phi}_{L,21}^{i_0 j_0 \ldots i_L j_L} & \check{\Phi}_{L,22}^{i_0 j_0 \ldots i_L j_L} \end{bmatrix},$$

$$\begin{bmatrix} \mathbf{A}(v_t, v_t) & \mathbf{B}(v_t) \\ \mathbf{C}(v_t) & \mathbf{D}(v_t) \end{bmatrix} \triangleq \begin{bmatrix} M & 0 \\ 0 & I \end{bmatrix} \begin{bmatrix} \widehat{A}(v_t, v_t) & \widehat{B}(v_t) \\ \widehat{C}(v_t) & \widehat{D}(v_t) \end{bmatrix}$$

$$\times \begin{bmatrix} N & 0 \\ C_y(v_t)X & I \end{bmatrix} + \begin{bmatrix} YA(v_t)X & 0 \\ 0 & 0 \end{bmatrix}. \tag{3.44}$$

Then it is easy to show that

$$GS = \begin{bmatrix} X & I \\ N & 0 \end{bmatrix}, \tag{3.45}$$

$$S^T \left\{ G + G^T \right\} S = \begin{bmatrix} X + X^T & I + Z^T \\ I + Z & Y + Y^T \end{bmatrix}, \tag{3.46}$$

$$\check{C}(v_t)GS = \begin{bmatrix} -C_z(v_t)X + \mathbf{C}(v_t) & -C_z(v_t) + \mathbf{D}(v_t)C_y(v_t) \end{bmatrix}, \tag{3.47}$$

$$\bar{E}^a(v_t)GS = \begin{bmatrix} E^a(v_t)X & E^a(v_t) \\ 0 & 0 \end{bmatrix}, \tag{3.48}$$

$$\check{C}(v_t)GS = \begin{bmatrix} -C_z(v_t)X + \mathbf{C}(v_t) & -C_z(v_t) + \mathbf{D}(v_t)C_y(v_t) \end{bmatrix}, \quad (3.49)$$

$$\bar{H}^T S = \begin{bmatrix} H^T & H^T Y^T \\ 0 & 0 \end{bmatrix}, \quad S^T \check{B}(v_t) = \begin{bmatrix} D_x(v_t) \\ Y D_x(v_t) + \mathbf{B}(v_t)D_y(v_t) \end{bmatrix},$$
$$(3.50)$$

$$S^T \bar{A}(v_t)GS = \begin{bmatrix} A(v_t)X & A(v_t) \\ \mathbf{A}(v_t, v_t) & YA(v_t) + \mathbf{B}(v_t)C_y(v_t) \end{bmatrix}, \quad (3.51)$$

$$\check{D}(v_t) = \mathbf{D}(v_t)D_y(v_t).$$

Pre- and postmultiplying (3.38) by $diag\{S^T, I, I, S^T, I, I\}$ and its transpose, respectively, (3.38) holds if and only if

$$\left[\begin{array}{ccc}
\bar{\Phi}_{k-1,11} - \bar{\mathbf{P}}_{k-1,11}(v_{t+k-1}) & \bullet & \bullet \\
\bar{\Phi}_{k-1,21} - \bar{\mathbf{P}}_{k-1,21}(v_{t+k-1}) & \bar{\Phi}_{k-1,22} - \bar{\mathbf{P}}_{k-1,22}(v_{t+k-1}) & \bullet \\
0_{dn} & 0_{dn} & -\gamma I_d \\
-C_z(v_{t+k-1})X + \mathbf{C}(v_{t+k-1}) & -C_z(v_{t+k-1}) + \mathbf{D}(v_{t+k-1})C_y(v_{t+k-1}) & \mathbf{D}(v_{t+k-1})D_y(v_{t+k-1}) \\
A(v_{t+k-1})X & A(v_{t+k-1}) & D_x(v_{t+k-1}) \\
\mathbf{A}(v_{t+k-1}, v_{t+k-1}) & YA(v_{t+k-1}) + \mathbf{B}(v_{t+k-1})C_y(v_{t+k-1}) & YD_x(v_{t+k-1}) + \mathbf{B}(v_{t+k-1})D_y(v_{t+k-1}) \\
E^a(v_{t+k-1})X & E^a(v_{t+k-1}) & 0_{nd} \\
0_{nn} & 0_{nn} & 0_{nd} \\
0_{nn} & 0_{nn} & 0_{nd} \\
0_{nn} & 0_{nn} & 0_{nd}
\end{array} \right.$$

$$\left. \begin{array}{cccccc}
\bullet & \bullet & \bullet & \bullet & \bullet & \bullet \\
\bullet & \bullet & \bullet & \bullet & \bullet & \bullet \\
\bullet & \bullet & \bullet & \bullet & \bullet & \bullet \\
-\gamma I_p & \bullet & \bullet & \bullet & \bullet & \bullet \\
0_{np} & -X - X^T + \bar{\mathbf{P}}_{k,11}(v_{t+k}) & \bullet & \bullet & \bullet & \bullet \\
0_{np} & -I_n - Z + \bar{\mathbf{P}}_{k,21}(v_{t+k}) & -Y - Y^T + \bar{\mathbf{P}}_{k,22}(v_{t+k}) & \bullet & \bullet & \bullet \\
0_{np} & 0_{nn} & 0_{nn} & -\rho_1 I_n & \bullet & \bullet \\
0_{np} & 0_{nn} & 0_{nn} & 0_{nn} & -\rho_1 I_n & \bullet \\
0_{np} & H^T & H^T Y^T & 0_{nn} & 0_{nn} & -\rho_1^{-1} I_n & \bullet \\
0_{np} & 0_{nn} & 0_{nn} & 0_{nn} & 0_{nn} & 0_{nn} & -\rho_1^{-1} I_n
\end{array} \right] < 0.$$
$$(3.52)$$

According to Lemma 2.1 in Chapter 2, if (3.12) for $g = 1$ holds, then so does (3.52), and hence $\check{\chi}_1 < 0$.

So, from (3.21) we have

$$\Delta_k \check{V}(\check{x}_t) < \check{x}_{t+k-1}^T \check{\Pi}_{k-1} \check{x}_{t+k-1} + \Delta_{k-2} \check{V}(\check{x}_t)$$

$$- \frac{1}{\gamma} e_{t+k-1}^T e_{t+k-1} + \gamma w_{t+k-1}^T w_{t+k-1} \quad (3.53)$$

$$- \check{x}_t^T \left(\sum_{g=k-1}^{k} \check{P}_g(v_t) \right) \check{x}_t.$$

Step 2: Adding and subtracting $\check{x}_{t+k-2}^T \check{\Pi}_{k-2} \check{x}_{t+k-2} + \frac{1}{\gamma} e_{t+k-2}^T e_{t+k-2} - \gamma w_{t+k-2}^T w_{t+k-2}$ to and from (3.53) and rearranging its terms give

$$
\begin{aligned}
\Delta_k \check{V}(\check{x}_t) &< \check{x}_{t+k-1}^T \check{\Pi}_{k-1} \check{x}_{t+k-1} + \check{x}_{t+k-2}^T \check{P}_{k-2}(v_{t+k-2}) \check{x}_{t+k-2} \\
&\quad - \check{x}_{t+k-2}^T \check{\Pi}_{k-2} \check{x}_{t+k-2} + \frac{1}{\gamma} e_{t+k-2}^T e_{t+k-2} - \gamma w_{t+k-2}^T w_{t+k-2} \quad (3.54) \\
&\quad + \check{x}_{t+k-2}^T \check{\Pi}_{k-2} \check{x}_{t+k-2} + \Delta_{k-3} \check{V}(\check{x}_t) \\
&\quad - \sum_{g=k-2}^{k-1} \left(\frac{1}{\gamma} e_{t+g}^T e_{t+g} - \gamma w_{t+g}^T w_{t+g} \right) - \check{x}_t^T \left(\sum_{g=k-2}^{k} \check{P}_g(v_t) \right) \check{x}_t.
\end{aligned}
$$

Using the system equation (3.4), we have $\check{x}_{t+k-1} = \check{A}(v_{t+k-2}) \check{x}_{t+k-2} + \check{B}(v_{t+k-2}) w_{t+k-2}$ and $e_{t+k-2} = \check{C}(v_{t+k-2}) \check{x}_{t+k-2} + \check{D}(v_{t+k-2}) w_{t+k-2}$ and substituting into (3.54), we obtain

$$
\begin{aligned}
\Delta_k \check{V}(\check{x}_t) &< \zeta_{t+k-2}^T \check{X}_2 \zeta_{t+k-2} + \check{x}_{t+k-2}^T \check{\Pi}_{k-2} \check{x}_{t+k-2} + \Delta_{k-3} \check{V}(\check{x}_t) \\
&\quad - \sum_{g=k-2}^{k-1} \left(\frac{1}{\gamma} e_{t+g}^T e_{t+g} - \gamma w_{t+g}^T w_{t+g} \right) \quad (3.55) \\
&\quad - \check{x}_t^T \left(\sum_{g=k-2}^{k} \check{P}_g(v_t) \right) \check{x}_t,
\end{aligned}
$$

where

$$
\check{X}_2 = \begin{bmatrix} \check{X}_2^{11} & \check{X}_2^{12} \\ \bullet & \check{X}_2^{22} \end{bmatrix}, \quad (3.56)
$$

$$
\begin{aligned}
\check{X}_2^{11} &= \check{A}^T(v_{t+k-2}) \check{\Pi}_{k-1} \check{A}(v_{t+k-2}) - \check{\Pi}_{k-2} + \check{P}_{k-2}(v_{t+k-2}) \\
&\quad + \frac{1}{\gamma} \check{C}^T(v_{t+k-2}) \check{C}(v_{t+k-2}), \\
\check{X}_2^{12} &= \check{A}^T(v_{t+k-2}) \check{\Pi}_{k-1} \check{B}(v_{t+k-2}) + \frac{1}{\gamma} \check{C}^T(v_{t+k-2}) \check{D}(v_{t+k-2}), \quad (3.57) \\
\check{X}_2^{22} &= \check{B}(v_{t+k-2}) \check{\Pi}_{k-1} \check{B}(v_{t+k-2}) - \gamma I_d + \frac{1}{\gamma} \check{D}^T(v_{t+k-2}) \check{D}(v_{t+k-2}).
\end{aligned}
$$

Following the same procedure of *Step 1*, (3.12) for $g = 2$ leads to $\check{X}_2 < 0$, and we obtain

$$
\begin{aligned}
\Delta_k \check{V}(\check{x}_t) &< \check{x}_{t+k-2}^T \check{\Pi}_{k-2} \check{x}_{t+k-2} + \Delta_{k-3} \check{V}(\check{x}_t) \\
&\quad - \sum_{g=k-2}^{k-1} \left(\frac{1}{\gamma} e_{t+g}^T e_{t+g} - \gamma w_{t+g}^T w_{t+g} \right) \quad (3.58) \\
&\quad - \check{x}_t^T \left(\sum_{g=k-2}^{k} \check{P}_g(v_t) \right) \check{x}_t.
\end{aligned}
$$

Repeating these steps $k - 3$ times, we arrive at

$$
\begin{aligned}
\Delta_k \check{V}(\check{x}_t) &< \check{x}_{t+1}^T \check{\Pi}_1 \check{x}_{t+1} - \check{x}_t^T \left(\sum_{g=1}^{k} \check{P}_g(v_t) \right) \check{x}_t \\
&\quad - \sum_{g=1}^{k-1} \left(\frac{1}{\gamma} e_{t+g}^T e_{t+g} - \gamma w_{t+g}^T w_{t+g} \right). \quad (3.59)
\end{aligned}
$$

Step k: Adding and subtracting $\frac{1}{\gamma}e_t^T e_t - \gamma w_t^T w_t$ to and from (3.59), using the system equation (3.4) again $\check{x}_{t+1} = \check{A}(v_t)\check{x}_t + \check{B}(v_t)w_t$ and $e_t = \check{C}(v_t)\check{x}_t + \check{D}(v_t)w_t$, substituting into (3.59), and rearranging its terms give

$$\Delta_k \check{V}(\check{x}_t) < \zeta_t^T \check{\chi}_k \zeta_t - \sum_{g=0}^{k-1}\left(\frac{1}{\gamma}e_{t+g}^T e_{t+g} - \gamma w_{t+g}^T w_{t+g}\right), \qquad (3.60)$$

where

$$\check{\chi}_k = \begin{bmatrix} \check{\chi}_k^{11} & \check{\chi}_k^{12} \\ \bullet & \check{\chi}_k^{22} \end{bmatrix}, \qquad (3.61)$$

$$
\begin{aligned}
\check{\chi}_k^{11} &= \check{A}^T(v_t)\check{\Pi}_1 \check{A}(v_t) - \sum_{g=1}^k \check{P}_g(v_t) + \frac{1}{\gamma}\check{C}^T(v_t)\check{C}(v_t), \\
\check{\chi}_k^{12} &= \check{A}^T(v_t)\check{\Pi}_1 \check{B}(v_t) + \frac{1}{\gamma}\check{C}^T(v_t)\check{D}(v_t), \qquad\qquad (3.62) \\
\check{\chi}_k^{22} &= \check{B}(v_t)\check{\Pi}_1 \check{B}(v_t) - \gamma I_d + \frac{1}{\gamma}\check{D}^T(v_t)\check{D}(v_t).
\end{aligned}
$$

Similarly, it can be shown that (3.12) for $g = k$ implies that $\check{\chi}_k < 0$. Hence, from (3.60) we have

$$\Delta_k \check{V}(\check{x}_t) + \sum_{g=0}^{k-1}\left(\frac{1}{\gamma}e_{t+g}^T e_{t+g} - \gamma w_{t+g}^T w_{t+g}\right) < 0. \qquad (3.63)$$

Taking the summation from $t = -k$ to $t = \infty$ on both sides of (3.63) and using (3.15)–(3.17), we have

$$\sum_{t=-k}^{\infty}\sum_{g=0}^{k-1}\left(\check{V}_{g+1}(\check{x}_{t+g+1}) - \check{V}_{g+1}(\check{x}_t) + \frac{1}{\gamma}e_{t+g}^T e_{t+g} - \gamma w_{t+g}^T w_{t+g}\right) < 0. \quad (3.64)$$

Expanding summations and rearranging terms in (3.64), we obtain

$$
\begin{aligned}
\sum_{j=0}^{k-1}&\left(\sum_{g=j}^{k-1}\check{V}_{g+1}(\check{x}_\infty) - \check{V}_{g+1}(\check{x}_{g-k})\right) \\
&+ \sum_{g=0}^{k-1}\left(\frac{1}{\gamma}\sum_{t=-k}^{\infty}e_{t+g}^T e_{t+g} - \gamma \sum_{t=-k}^{\infty}w_{t+g}^T w_{t+g}\right) < 0. \qquad (3.65)
\end{aligned}
$$

Under zero initial conditions ($\check{x}_0 = \check{x}_{-1} = \ldots = \check{x}_{-k} = 0$), for nonzero $w_t \in L_2[0, \infty)$, we have $\check{V}_{g+1}(\check{x}_{g-k}) = 0$, so that (3.65) is simplified to

$$\sum_{j=0}^{k-1}\left(\sum_{g=j}^{k-1}\check{V}_{g+1}(\check{x}_\infty)\right) + k\left(\frac{1}{\gamma}\sum_{t=0}^{\infty}e_t^T e_t - \gamma \sum_{t=0}^{\infty}w_t^T w_t\right) < 0. \qquad (3.66)$$

On the other hand, we know that $\mu_g(v_t) \geq 0$ for all $g \in \mathbb{K}_{1,r}$ and $\sum_{g=1}^r \mu_g(v_t) = 1$. So, if (3.11) holds, then $\sum_{g=j}^k \bar{\mathbf{P}}_g(v_t) > 0$ for all $j \in \mathbb{K}_{1,k}$. From (3.16) we can show that $\sum_{g=j}^k \check{P}_g(v_t) > 0$ for all $j \in \mathbb{K}_{1,k}$, so that we can conclude that

$$\sum_{g=j}^k \check{V}_g(\check{x}_t) > 0 \ \forall \ j \in \mathbb{K}_{1,k}. \tag{3.67}$$

From (3.67), $\sum_{j=0}^{k-1} \left(\sum_{g=j}^{k-1} \check{V}_{g+1}(\check{x}_\infty) \right) > 0$, so that (3.66) is reduced to $\frac{1}{\gamma} \sum_{t=0}^\infty e_t^T e_t - \gamma \sum_{t=0}^\infty w_t^T w_t < 0$, which implies that H_∞ performance in (3.9) is fulfilled.

Next, we show that the filtering error system (3.4) is globally asymptotically stable when the disturbance $w_t = 0$. In (3.63), if $w_t = 0$, then $\Delta_k \check{V}(\check{x}_t) + \frac{1}{\gamma} \sum_{g=0}^k e_{t+g}^T e_{t+g} < 0$. This implies that $\Delta_k \check{V}(\check{x}_t) < 0$. Therefore, according to Lemma 2.2 in Chapter 2, the filtering error system (3.4) is globally asymptotically stable.

It is necessary to check the existence of N^{-1} and M^{-1}. Note that if (3.12) for $g = k$ holds, then

$$\begin{bmatrix} -X - X^T + \bar{\mathbf{P}}_{k,11}^{ik} & \bullet \\ -I_n - Z + \bar{\mathbf{P}}_{k,21}^{ik} & -Y - Y^T + \bar{\mathbf{P}}_{k,22}^{ik} \end{bmatrix} < 0, \tag{3.68}$$

which in turns implies that

$$\begin{bmatrix} -X - X^T + \bar{\mathbf{P}}_{k,11}(v_t) & \bullet \\ -I_n - Z + \bar{\mathbf{P}}_{k,21}(v_t) & -Y - Y^T + \bar{\mathbf{P}}_{k,22}(v_t) \end{bmatrix} < 0. \tag{3.69}$$

We know that $\sum_{g=j}^k \bar{\mathbf{P}}_g(v_t) > 0$ all $j \in \mathbb{K}_{1,k}$. Letting $j = k$, we have

$$\bar{\mathbf{P}}_k(v_t) = \begin{bmatrix} \bar{\mathbf{P}}_{k,11}(v_t) & \bullet \\ \bar{\mathbf{P}}_{k,21}(v_t) & \bar{\mathbf{P}}_{k,22}(v_t) \end{bmatrix} > 0. \tag{3.70}$$

From (3.69) and (3.70) we have

$$\begin{bmatrix} X + X^T & \bullet \\ I_n + Z & Y + Y^T \end{bmatrix} > \begin{bmatrix} \bar{\mathbf{P}}_{k,11}(v_t) & \bullet \\ \bar{\mathbf{P}}_{k,21}(v_t) & \bar{\mathbf{P}}_{k,22}(v_t) \end{bmatrix} > 0. \tag{3.71}$$

Hence, X and Y are nonsingular. Pre- and postmultiplying (3.71) by $\begin{bmatrix} X^{-T} & I_n \end{bmatrix}$ and its transpose, respectively, gives $(Z - YX)X^{-1} + X^{-T}(Z - YX)^T < 0$, which implies that $Z - YX$ is nonsingular. So there exist nonsingular matrices N and M such that $MN = Z - YX$. This completes the proof. □

Remark 3.1. Theorem 3.1 involves kr LMIs of size $(2n)$, n_L LMIs of size $(8n + d + p)$, and $(n_p + n_f + n_\Pi + 3n^2 + 1)$ scalar variables, where $n_p = rkn(2n + 1)$, $n_f = r\left(n^2 + np + np_y + pp_y\right)$, and

$$n_L = \begin{cases} r^{2k+1} + \sum_{i=0}^{k-2}\left(r^{4+2i}\right), & k > 1, \\ r^4, & k = 1, \end{cases} \tag{3.72}$$

$$n_\Pi = \begin{cases} n(2n + 1)\sum_{i=0}^{k-2}\left(r^{4+2i}\right), & k > 1, \\ 0, & k = 1. \end{cases} \tag{3.73}$$

Remark 3.2. It is important to mention that the goal of using the theorem is to design a filter with H_∞ performance when there is no solution using the standard Lyapunov theory. If we have no any feasible filter using LMI tool-box in Matlab software, then we try for $k = 2$. If we cannot find any solution again, we try for $k = 3$, and so on.

3.4 SIMULATION RESULTS

Consider the two-rule model of the stable T-S fuzzy system with uncertainties in the form of (3.2), where $A_1 = \begin{bmatrix} 0.1 & 0.35 \\ -0.42 & -0.07 \end{bmatrix}$, $A_2 = \begin{bmatrix} 0.8 & 0.8617 \\ -0.0172 & 0.8103 \end{bmatrix}$, $D_{x,1} = \begin{bmatrix} 0.091 \\ 0.181 \end{bmatrix}$, $D_{x,2} = \begin{bmatrix} 0.093 \\ 0.181 \end{bmatrix}$, $C_{y,1} = C_{y,2} = \begin{bmatrix} 1 & 1 \end{bmatrix}$, $D_{y,1} = D_{y,2} = 0$, $C_{z,1} = C_{z,2} = \begin{bmatrix} 1 & 0 \end{bmatrix}$, $H = 0.1I_2$, $E_1^a = E_2^a = \begin{bmatrix} 0.05 & 0 \\ 0 & 0.01 \end{bmatrix}$.

The robust H_∞ output feedback filtering is designed in the following two cases.

Case 1: The approach proposed in Theorem 3.1.
Case 2: The k-sample variation approach [10–12,6,7].

The parameters $\hat{\rho}_1$, $\hat{\rho}_2$, and $\hat{\rho}_3$ are all selected to be 0.1. The optimal H_∞ performance $\gamma_{\min}$ is computed for $k = 1, 2$, and 3. Table 3.1 shows the optimal $\gamma_{\min}$ for different values of k, number of LMIs, and scalar variables. As expected, $\gamma_{\min}$ decreases as k increases. Moreover, Theorem 3.1

TABLE 3.1 $\gamma_{\min}$, number of LMIs, and scalar variables.

	$k = 1$	$k = 2$	$k = 3$
Theorem 3.1	$\gamma_{\min} = 2.5477$ 2 LMIs of 4×4	$\gamma_{\min} = 1.6615$ 4 LMIs of 4×4 48 LMIs of 18×18 231 variables	$\gamma_{\min} = 1.4529$ 6 LMIs of 4×4 208 LMIs of 18×18 891 variables
k-sample variation [10–12,6,7]	8 LMIs of 18×18 51 variables	$\gamma_{\min} = 2.2644$ 2 LMIs of 4×4 48 LMIs of 18×18 211 variables	$\gamma_{\min} = 1.6004$ 2 LMIs of 4×4 208 LMIs of 18×18 851 variables

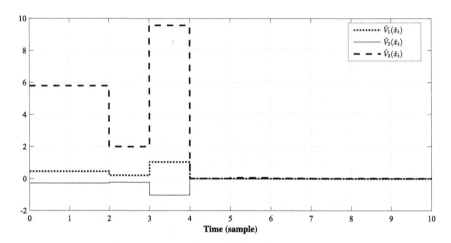

FIGURE 3.1 $\hat{V}_1(\check{x}_t)$, $\hat{V}_2(\check{x}_t)$, and $\hat{V}_3(\check{x}_t)$ in Theorem 3.1 with $k = 3$.

always shows better results than the k-sample variation approach. Fig. 3.1 shows $\hat{V}_1(\check{x}_t)$, $\hat{V}_2(\check{x}_t)$, and $\hat{V}_3(\check{x}_t)$ when $k = 3$ and initial conditions are $x_t = (0.1, -1)$ and $\bar{x}_t = (0, 0)$. From the figure we can observe that the stability of the filtering error system is ensured; however, they are decreasing nonmonotonically. Summing up, this example shows that the developed approach indeed offers less conservative results than standard Lyapunov theory and even the k-sample variation approach.

The designed filter using Theorem 3.1 for $k = 3$ is associated with the system. For simulations, $\upsilon_1(t) = x_1(t)$ is the premise variable, and $\mu_1(\upsilon_1(t)) = 0.5 + 0.5\cos(\upsilon_1(t))$ and $\mu_2(\upsilon_1(t)) = 1 - \mu_1(\upsilon_1(t))$, with sampling time 0.01. The obtained H_∞ performance is 0.0025, as shown in Fig. 3.2, which is the ratio of the regulated output z_t energy to the disturbance energy $w_t = 0.1\sin(t)$. This implies that the L_2 gain from the disturbance to the regulated output is no greater than $\gamma_{\min} = 0.0025$ and is much less than the prescribed value of $\gamma_{\min} = 1.4529$.

Estimation errors for both case k-sample variation and methodology developed in Theorem 3.1 for $k = 3$ are shown in Fig. 3.3.

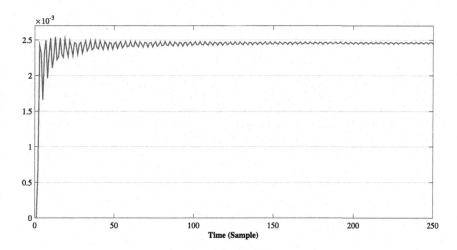

FIGURE 3.2 The ratio of the controlled output energy to the disturbance γ_{min} for $k = 3$.

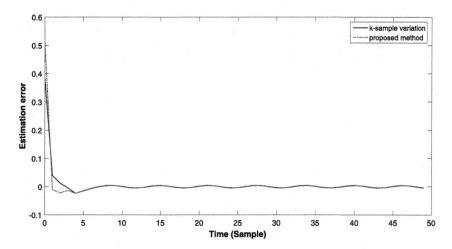

FIGURE 3.3 Estimation error signal for $k = 3$ in both cases.

3.5 CONCLUSIONS

In this chapter, the NLF has been developed to design robust full-order robust H_∞ filtering for uncertain Takagi–Sugeno (T-S) fuzzy systems. Some more slack matrix variables have been introduced to enlarge the design space. Based on that, sufficient conditions for the existence of a filter have been given in terms of linear matrix inequalities (LMIs). It has been shown that the conservatism is reduced, and the H_∞ performance is improved via a numerical example.

References

[1] T. Takagi, M. Sugeno, Fuzzy identification of systems and its applications to modeling and control, IEEE Transactions on Systems, Man and Cybernetics SMC-15 (1) (Jan 1985) 116–132.

[2] G. Feng, A survey on analysis and design of model-based fuzzy control systems, IEEE Transactions on Fuzzy Systems 14 (5) (Oct 2006) 676–697.

[3] G. Feng, Stability analysis of discrete-time fuzzy dynamic systems based on piecewise Lyapunov functions, IEEE Transactions on Fuzzy Systems 12 (1) (Feb 2004) 22–28.

[4] H.-N. Wu, K.-Y. Cai, H_2 guaranteed cost fuzzy control design for discrete-time nonlinear systems with parameter uncertainty, Automatica 42 (7) (July 2006) 1183–1188.

[5] S.H. Kim, Improved approach to robust $\mathcal{H}_\infty$ stabilization of discrete-time T-S fuzzy systems with time-varying delays, IEEE Transactions on Fuzzy Systems 18 (5) (2010) 1008–1015.

[6] A. Kruszewski, R. Wang, T. Guerra, Nonquadratic stabilization conditions for a class of uncertain nonlinear discrete time T-S fuzzy models: a new approach, IEEE Transactions on Automatic Control 53 (2) (March 2008) 606–611.

[7] Y.-J. Chen, M. Tanaka, K. Inoue, H. Ohtake, K. Tanaka, T. Guerra, A. Kruszewski, H. Wang, A nonmonotonically decreasing relaxation approach of Lyapunov functions to guaranteed cost control for discrete fuzzy systems, IET Control Theory & Applications 8 (16) (2014) 1716–1722.

[8] S.F. Derakhshan, A. Fatehi, Non-monotonic Lyapunov functions for stability analysis and stabilization of discrete time Takagi–Sugeno fuzzy systems, International Journal of Innovative Computing, Information & Control (2014) 1567–1586.

[9] A. Ahmadi, P. Parrilo, Non-monotonic Lyapunov functions for stability of discrete time nonlinear and switched systems, in: 47th IEEE Conference on Decision and Control, 2008 (CDC 2008), Dec 2008, pp. 614–621.

[10] S.F. Derakhshan, A. Fatehi, Non-monotonic robust H_2 fuzzy observer-based control for discrete time nonlinear systems with parametric uncertainties, International Journal of Systems Science 46 (12) (2015) 2134–2149.

[11] S. Derakhshan, A. Fatehi, M. Sharabiany, Nonmonotonic observer-based fuzzy controller designs for discrete time T-S fuzzy systems via LMI, IEEE Transactions on Cybernetics (99) (April 2014).

[12] T. Guerra, A. Kruszewski, M. Bernal, Control law proposition for the stabilization of discrete Takagi–Sugeno models, IEEE Transactions on Fuzzy Systems 17 (3) (June 2009) 724–731.

[13] M.C.D. Oliveira, J.C. Geromel, J. Berussou, Extended H_2 and H norm characterizations and controller parametrizations for discrete-time systems, International Journal of Control 75 (9) (2002) 666–679.

[14] L. Xie, Output feedback H_∞ control of system with parameter uncertainty, International Journal of Control 63 (4) (1996) 741–750.

4

Robust H_∞ Output Feedback Control for T-S Fuzzy Systems: A Nonmonotonic Approach

4.1 INTRODUCTION

In many practical systems, the measured signals are often subjected to external disturbance (noise). Consequently, it is desirable to mitigate its effects using one of robust control techniques. Robust H_∞ control has attracted great attention from both scientists and engineers as an effective tool to tackle such problem. Robust H_∞ output feedback control is often designed to minimize the ratio of the controlled output energy to the disturbance energy.

It is well known that most of practical systems are inherently nonlinear, and one of the widely used techniques to approximate nonlinear systems is the Takagi–Sugeno (T-S) fuzzy model [1]. T-S fuzzy model is a combination of linear models, which are connected together by nonlinear scalar membership functions. Such a minimum complexity feature of T-S fuzzy model has attracted large attention for H_∞ control problem. A comprehensive review of various types of controllers, which have been designed for T-S fuzzy systems, can be found in [2] and references therein. To mention a few, robust H_∞ state feedback and output feedback controllers are designed in [3] and [4–6], respectively.

Generally speaking, the stability analysis of T-S model and subsequently controller design is often carried out through the standard Lyapunov method. This inherently leads to some inequalities, in which the nonlinear membership functions are excluded from final conditions. As a result, the obtained conditions, which are often constructed in the form of linear matrix inequalities (LMIs) [7], become only sufficient (conservative). In the literature, two main approaches have been presented to mitigate the demerits of this conservatism. The first approach focuses on judiciously

61

choosing a suitable Lyapunov function to reduce the conservatism, e.g., piecewise [8], fuzzy [9], nonquadratic LFs [10], and combinations of these. The second approach, which is the main focus of this study, reduces the conservatism by relaxing the monotonic decrease of LF in discrete-time systems. This approach is alternately referred to as the nonmonotonic approach [11–16]. In other words, LFs in this approach are not required anymore to decrease in each successive step (i.e., $V_{t+1} < V_t$). Instead, it allows small local growth, but it ultimately converges to zero [11]. Recently, some works have been made using the nonmonotonic approach. In [17, 16,15], the nonmonotonic decrease of LF is based on only 2-sample variations (i.e., $V_{t+2} < V_t$). This nonmonotonic approach has been shown to be less conservative than the common quadratic LF [18] and even piecewise LF [19]. A more general NLF with k-sample variations (i.e., $V_{t+k} < V_t$) has been introduced in [12–14]. In [12] and [13], a robust state feedback controller for uncertain discrete-time T-S fuzzy systems has been designed. Moreover, the same approach is considered for guaranteed cost control design in [14].

Robust H_∞ control turns out to be more difficult to solve when compared to the aforecited control problems using NLFs. Thus, it remains challenging and needs to be solved. This chapter extends k-sample variations [12–16] to new NLF approach. This new NLF approach introduces more decision variables and inequalities to further reduce the conservatism. Then the design of robust H_∞ controller, which is the main objective of this chapter, is investigated using the extended NLF approach. Finally, sufficient conditions for the existence of robust H_∞ controller are derived in terms of LMIs.

The rest of the chapter is organized as follows. Section 4.2 briefly describes the system and problem formulation. The main results are presented in Section 4.3. The effectiveness of the developed controller design is demonstrated in Section 4.4. Conclusions are drawn in Section 4.5.

Notation: Throughout the chapter, $(\cdot)^T$ and $\bullet$ denote, respectively, the transpose and the symmetric block in a symmetric matrix. The identity matrix of size n is denoted by I_n, and 0_{nm} represents the zero matrix of size $n \times m$; $x_t \triangleq x(t)$ and $\mathbb{K}_{r_1,r_L} \triangleq \{r_1, \ldots, r_L\}$, where $t, r_1, \ldots, r_L \in \mathbb{N}$; $X(v_t) \triangleq \sum_{i=1}^{r} \mu_i(v_t)X_i$ with $\mu_i(v_t) \geq 0$ and $\sum_{i=1}^{r} \mu_i(v_t) = 1$.

4.2 PRELIMINARIES AND PROBLEM FORMULATION

This chapter considers the following uncertain discrete-time T-S fuzzy system:

Plant Rule i: IF $v_{1,t}$ is M_1^i AND $\cdots$ AND $v_{\sigma,t}$ is M_σ^i, THEN

$$
\begin{aligned}
x_{t+1} &= (A_i + \Delta A_i)x_t + \left(B_{x,i} + \Delta B_{x,i}\right)u_t + D_{x,i}w_t, \\
y_t &= C_{y,i}x_t + D_{y,i}w_t, \\
z_k &= C_{z,i}x_t + B_{z,i}u_t,
\end{aligned}
\tag{4.1}
$$

where $x_t \in \mathbb{R}^n$, $u_t \in \mathbb{R}^m$, $z_t \in \mathbb{R}^p$, $y_t \in \mathbb{R}^{p_y}$, and $w_t \in \mathbb{R}^d$ are the state, control, regulated output, measured output signals, and disturbance, respectively, $i \in \mathbb{K}_{1,r}$ denotes the ith fuzzy inference rule, r is the number of rules, $v_{\tau,t}$ are the premise variables ($v_t = [v_{1,t}, \ldots, v_{\sigma,t}]$), σ is the number of premise variables, and $M_1^i, \ldots, M_\sigma^i$ are membership functions. The matrices A_i, ΔA_i, $B_{x,i}$, $\Delta B_{x,i}$, $D_{x,i}$, $D_{y,i}$, $C_{y,i}$, $C_{z,i}$, and $B_{z,i}$ are of appropriate dimensions. By using a center-average defuzzifier, product inference, and singleton fuzzifier, the uncertain T-S fuzzy system (4.1) can be expressed as follows:

$$
\begin{aligned}
x_{t+1} &= (A(v_t) + \Delta A(v_t))x_t + (B_x(v_t) + \Delta B_x(v_t))u_t + D_x(v_t)w_t, \\
y_t &= C_y(v_t)x_t + D_y(v_t)w_t, \\
z_t &= C_z(v_t)x_t + B_z(v_t)u_t,
\end{aligned}
\tag{4.2}
$$

where $\mu_i(v_t) = \frac{\prod_{\tau=1}^\sigma M_\tau^i(v_{\tau,t})}{\sum_{\ell=1}^r \prod_{\tau=1}^\sigma M_\tau^\ell(v_{\tau,t})}$ with $\mu_i(v_t) \geq 0$ and $\sum_{i=1}^r \mu_i(v_t) = 1$. The system uncertainties are assumed to be $[\Delta A(v_t)\ \Delta B_x(v_t)] = HF(x_t, t)[E^a(v_t)\ E^b(v_t)]$, where H, $E^a(v_t)$, and $E^b(v_t)$ are known matrix functions, which characterize the structure of the uncertainties, and $F(x_t, t)$ is an unknown nonlinear time-varying matrix function satisfying $F(x_t, t)^T F(x_t, t) \leq I$.

This chapter aims at solving the following parallel distributed controller (PDC) output feedback H_∞ problem.

Robust H_∞ output Feedback control Problem Formulation: Given a prescribed H_∞ performance γ, design a PDC output feedback controller

$$
\begin{aligned}
\hat{x}_{t+1} &= \sum_{j=1}^r \sum_{i=1}^r \mu_i(v_t)\mu_j(v_t)\left\{\hat{A}_{ij}\hat{x}_t + \hat{B}_j y_t\right\} \\
&= \hat{A}(v_t, v_t)\hat{x}_t + \hat{B}(v_t)y_t, \\
u_t &= \sum_{j=1}^r \mu_j(v_t)\left\{\hat{C}_j\hat{x}_t + \hat{D}_j y_t\right\} \\
&= \hat{C}(v_t)\hat{x}_t + \hat{D}(v_t)y_t,
\end{aligned}
\tag{4.3}
$$

where $\hat{x}_t \in \mathbb{R}^n$ is controller states, and $\hat{A}_{ij}$, $\hat{B}_j$, $\hat{C}_j$, and $\hat{D}_j$ are of appropriate dimensions. System (4.2) with (4.3) is asymptotically stable, and the following inequality holds:

$$\sum_{t=0}^{\infty} z_t^T z_t < \gamma \sum_{t=0}^{\infty} \left(w_t^T w_t \right). \tag{4.4}$$

The closed-loop system of the T-S fuzzy system (4.2) with the fuzzy controller (4.3) is given by

$$\begin{aligned} \tilde{x}_{t+1} &= \tilde{A}(v_t)\tilde{x}_t + \tilde{B}(v_t)w_t, \\ z_t &= \tilde{C}(v_t)\tilde{x}_t + \tilde{D}(v_t)w_t, \end{aligned} \tag{4.5}$$

where $\tilde{x}_t^T = \begin{bmatrix} x_t^T & \hat{x}_t^T \end{bmatrix}$,

$$\tilde{A}(v_t) = \begin{bmatrix} A(v_t) + B_x(v_t)\hat{D}(v_t)C_y(v_t) + \triangle A(v_t) + \triangle B_x(v_t)\hat{D}(v_t)C_y(v_t) \\ \hat{B}(v_t)C_y(v_t) \\ B_x(v_t)\hat{C}(v_t) + \triangle B_x(v_t)\hat{C}(v_t) \\ \hat{A}(v_t, v_t) \end{bmatrix}, \tag{4.6}$$

$$\tilde{C}(v_t) = \begin{bmatrix} C_z(v_t) + B_z(v_t)\hat{D}(v_t)C_y(v_t) & B_z(v_t)\hat{C}(v_t) \end{bmatrix}, \tag{4.7}$$

$$\tilde{B}(v_t) = \begin{bmatrix} B_x(v_t)\hat{D}(v_t)D_y(v_t) + \triangle B_x(v_t)\hat{D}(v_t)D_y(v_t) + D_x(v_t) \\ \hat{B}(v_t)D_y(v_t) \end{bmatrix}, \tag{4.8}$$

$$\tilde{D}(v_t) = B_z(v_t)\hat{D}(v_t)D_y(v_t). \tag{4.9}$$

4.3 MAIN RESULTS

The following theorem provides sufficient conditions in terms of LMIs for the existence and design of robust H_∞ output feedback controller (4.3) using an NLF.

Theorem 4.1. *Given a constant $\gamma > 0$, the closed-loop control system (4.5) is globally asymptotically stable with disturbance attenuation γ if there exist positive scalar parameters ρ_g, a set of symmetric matrices $\bar{P}_{g,11}^{ih}$, $\bar{P}_{g,22}^{ih}$, $\bar{\Pi}_{L,11}^{i_0 j_0 \ldots i_L j_L}$, $\bar{\Pi}_{L,22}^{i_0 j_0 \ldots i_L j_L}$, and matrices X, Y, Z, $\bar{\Pi}_{L,21}^{i_0 j_0 \ldots i_L j_L}$, $\bar{P}_{g,21}^{ih}$ for $h \in \mathbb{K}_{0,k-1}$, $i_0 \ldots i_k j_0 \ldots j_{k-1} \in \mathbb{K}_{1,r}$, $L \in \mathbb{K}_{1,k-1}$, and $g \in \mathbb{K}_{1,k}$ such that*

$$\sum_{g=j}^{k} \bar{P}_{g,i_h} > 0 \qquad \forall\, i_h \in \mathbb{K}_{1,r} \text{ and } j \in \mathbb{K}_{1,k} \tag{4.10}$$

$$
\begin{cases}
\Lambda_{g,i_{k-g}i_{k-g}} < 0, \\
\frac{1}{r-1}\Lambda_{g,i_{k-g}i_{k-g}} + \frac{1}{2}\left(\Lambda_{g,i_{k-g}j_{k-g}} + \Lambda_{g,j_{k-g}i_{k-g}}\right) < 0 \quad \forall\, i_{k-g} \neq j_{k-g},
\end{cases} \tag{4.11}
$$

where

$$
\bar{P}_{g,i_h} = \begin{bmatrix} \bar{P}^{i_h}_{g,11} & \bullet \\ \bar{P}^{i_h}_{g,21} & \bar{P}^{i_h}_{g,22} \end{bmatrix}, \tag{4.12}
$$

$$
\Lambda_{g,i_{k-g}j_{k-g}} =
\left[
\begin{array}{cc}
\Lambda^{11}_g & \bullet \\
\Lambda^{21}_g & \Lambda^{22}_g \\
0_{dn} & 0_{dn} \\
C_{z,i_{k-g}}X + B_{z,i_{k-g}}C_{j_{k-g}} & C_{z,i_{k-g}} + B_{z,i_{k-g}}\mathcal{D}_{j_{k-g}}C_{y,i_{k-g}} \\
A_{i_{k-g}}X + B_{x,i_{k-g}}C_{j_{k-g}} & A_{i_{k-g}} + B_{x,i_{k-g}}\mathcal{D}_{j_{k-g}}C_{y,i_{k-g}} \\
A_{i_{k-g},j_{k-g}} & YA_{i_{k-g}} + \mathcal{B}_{j_{k-g}}C_{y,i_{k-g}} \\
E^a_{i_{k-g}}X + E^b_{i_{k-g}}C_{j_{k-g}} & E^a_{i_{k-g}} + E^b_{i_{k-g}}\mathcal{D}_{j_{k-g}}C_{y,i_{k-g}} \\
0_{nn} & 0_{nn} \\
0_{nn} & 0_{nn} \\
0_{nn} & 0_{nn}
\end{array}
\right.
$$

$$
\left.
\begin{array}{cccccccc}
\bullet & \bullet & \bullet & \bullet & \bullet & \bullet & \bullet & \bullet \\
\bullet & \bullet & \bullet & \bullet & \bullet & \bullet & \bullet & \bullet \\
-\gamma I_d & \bullet & \bullet & \bullet & \bullet & \bullet & \bullet & \bullet \\
B_{z,i_{k-g}}\mathcal{D}_{j_{k-g}}D_{y,i_{k-g}} & -\gamma I_p & \bullet & \bullet & \bullet & \bullet & \bullet & \bullet \\
D_{x,i_{k-g}} + B_{x,i_{k-g}}\mathcal{D}_{j_{k-g}}D_{y,i_{k-g}} & 0_{np} & \Lambda^{55}_g & \bullet & \bullet & \bullet & \bullet & \bullet \\
YD_{x,i_{k-g}} + \mathcal{B}_{j_{k-g}}D_{y,i_{k-g}} & 0_{np} & \Lambda^{65}_g & \Lambda^{66}_g & \bullet & \bullet & \bullet & \bullet \\
E^b_{i_{k-g}}\mathcal{D}_{j_{k-g}}D_{y,i_{k-g}} & 0_{np} & 0_{nn} & 0_{nn} & -\rho_g I_n & \bullet & \bullet & \bullet \\
0_{nd} & 0_{np} & 0_{nn} & 0_{nn} & 0_{nn} & -\rho_g I_n & \bullet & \bullet \\
0_{nd} & 0_{np} & H^T & H^T Y^T & 0_{nn} & 0_{nn} & -\rho^{-1}_g I_n & \bullet \\
0_{nd} & 0_{np} & 0_{nn} & 0_{nn} & 0_{nn} & 0_{nn} & 0_{nn} & -\rho^{-1}_g I_n
\end{array}
\right], \tag{4.13}
$$

$$
\Lambda^{11}_1 = \bar{\Pi}^{i_0 j_0 \ldots i_{k-1} j_{k-1}}_{k-1,11} - \bar{P}^{i_{k-1}}_{k-1,11}, \quad \Lambda^{21}_1 = \bar{\Pi}^{i_0 j_0 \ldots i_{k-1} j_{k-1}}_{k-1,21} - \bar{P}^{i_{k-1}}_{k-1,21},
$$

$$
\Lambda^{22}_1 = \bar{\Pi}^{i_0 j_0 \ldots i_{k-1} j_{k-1}}_{k-1,22} - \bar{P}^{i_{k-1}}_{k-1,22}, \quad \Lambda^{55}_1 = -X - X^T + \bar{P}^{i_k}_{k,11},
$$

$$
\Lambda^{65}_1 = -I_n - Z + \bar{P}^{i_k}_{k,21}, \quad \Lambda^{66}_1 = -Y - Y^T + \bar{P}^{i_k}_{k,22},
$$

$$
\Lambda^{11}_s = \bar{\Pi}^{i_0 j_0 \ldots i_{k-s} j_{k-s}}_{k-s,11} - \bar{P}^{i_{k-s}}_{k-s,11}, \quad \Lambda^{21}_s = \bar{\Pi}^{i_0 j_0 \ldots i_{k-s} j_{k-s}}_{k-s,21} - \bar{P}^{i_{k-s}}_{k-s,21},
$$

$$
\Lambda^{22}_s = \bar{\Pi}^{i_0 j_0 \ldots i_{k-s} j_{k-s}}_{k-s,22} - \bar{P}^{i_{k-s}}_{k-s,22}, \quad \Lambda^{55}_s = -X - X^T + \bar{\Pi}^{i_0 j_0 \ldots i_{k-s+1} j_{k-s+1}}_{k-s+1,11},
$$

$$\Lambda_s^{65} = -I_n - Z + \bar{\Pi}_{k-s+1,21}^{i_0 j_0 \cdots i_{k-s+1} j_{k-s+1}},$$

$$\Lambda_s^{66} = -Y - Y^T + \bar{\Pi}_{k-s+1,22}^{i_0 j_0 \cdots i_{k-s+1} j_{k-s+1}}, \quad s \in \mathbb{K}_{2,k-1},$$

$$\Lambda_k^{11} = -\sum_{i=1}^{k} \bar{P}_{i,11}^{i_0}, \quad \Lambda_k^{21} = -\sum_{i=1}^{k} \bar{P}_{i,21}^{i_0},$$

$$\Lambda_k^{22} = -\sum_{i=1}^{k} \bar{P}_{i,22}^{i_0}, \quad \Lambda_k^{55} = -X - X^T + \bar{\Pi}_{1,11}^{i_0 j_0 i_1 j_1},$$

$$\Lambda_k^{65} = -I_n - Z + \bar{\Pi}_{1,21}^{i_0 j_0 i_1 j_1}, \quad \Lambda_k^{66} = -Y - Y^T + \bar{\Pi}_{1,22}^{i_0 j_0 i_1 j_1}.$$

The controller parameters $\hat{A}_{i_h j_h} \in \mathbb{R}^{n \times n}$, $\hat{B}_{i_h} \in \mathbb{R}^{n \times p_y}$, $\hat{C}_{i_h} \in \mathbb{R}^{m \times n}$, and $\hat{D}_{i_h} \in \mathbb{R}^{m \times p_y}$ in (4.3) are given as follows:

$$\begin{bmatrix} \hat{A}_{i_h j_h} & \hat{B}_{i_h} \\ \hat{C}_{i_h} & \hat{D}_{i_h} \end{bmatrix} = \begin{bmatrix} M^{-1} & -M^{-1} Y B_{i_h} \\ 0 & I \end{bmatrix} \begin{bmatrix} A_{i_h j_h} - Y A_{i_h} X & B_{i_h} \\ C_{i_h} & \mathcal{D}_{i_h} \end{bmatrix}$$

$$\times \begin{bmatrix} N^{-1} & 0 \\ -C_{y,i_h} X N^{-1} & I \end{bmatrix}.$$

Note that, the two nonsingular constant matrices $N \in \mathbb{R}^{n \times n}$ and $M \in \mathbb{R}^{n \times n}$ can always be obtained such that $MN = Z - YX$.

Proof. Consider the following function:

$$\hat{V}_g(\tilde{x}_t) \triangleq \tilde{x}_t^T \hat{P}_g(v_t) \tilde{x}_t \quad \forall g \in \mathbb{K}_{1,k}, \tag{4.14}$$

where

$$\hat{P}_g(v_t) = G^{-T} P_g(v_t) G^{-1}. \tag{4.15}$$

Define

$$\Delta_k \hat{V}(\tilde{x}_t) \triangleq \sum_{g=1}^{k} \left(\tilde{x}_{t+g}^T \hat{P}_g(v_{t+g}) \tilde{x}_{t+g} - \tilde{x}_t^T \hat{P}_g(v_t) \tilde{x}_t \right), \tag{4.16}$$

$$\Pi_L \triangleq \sum_{i_L=1}^{r} \sum_{j_L=1}^{r} \cdots \sum_{i_0=1}^{r} \sum_{j_0=1}^{r} \mu_{i_0}(v_t) \mu_{j_0}(v_t) \cdots \mu_{i_L}(v_{t+L}) \mu_{j_L}(v_{t+L}) \tilde{\Pi}_{L,i_0 j_0 \cdots i_L j_L}, \tag{4.17}$$

$$\hat{\Pi}_L \triangleq G^{-T} \Pi_L G^{-1}. \tag{4.18}$$

As in the last theorem, we need to follow k steps to show that the theorem conditions satisfy H_∞ performance and stability of the closed-loop system. There are some important differences. For example, in the gth step, we add and subtract $\tilde{x}_{t+k-g}^T \hat{\Pi}_{k-g} \tilde{x}_{t+k-g} + \frac{1}{\gamma} z_{t+k-g}^T z_{t+k-g} - \gamma w_{t+k-g}^T w_{t+k-g}$, and the nonlinear transformation (change of variables) will be as follows:

$$\begin{bmatrix} \mathcal{A}(v_t, v_t) & \mathcal{B}(v_t) \\ \mathcal{C}(v_t) & \mathcal{D}(v_t) \end{bmatrix} \triangleq \begin{bmatrix} M & YB_x(v_t) \\ 0 & I \end{bmatrix} \begin{bmatrix} \hat{A}(v_t, v_t) & \hat{B}(v_t) \\ \hat{C}(v_t) & \hat{D}(v_t) \end{bmatrix}$$
$$\times \begin{bmatrix} N & 0 \\ C_y(v_t)X & I \end{bmatrix} + \begin{bmatrix} YA(v_t)X & 0 \\ 0 & 0 \end{bmatrix}. \tag{4.19}$$

The rest of the proof is the same.

Consider the function

$$\hat{V}_g(\tilde{x}_t) \triangleq \tilde{x}_t^T \hat{P}_g(v_t)\tilde{x}_t \quad \forall g \in \mathbb{K}_{1,k}, \tag{4.20}$$

where

$$\hat{P}_g(v_t) = G^{-T} P_g(v_t) G^{-1}. \tag{4.21}$$

Define

$$\Delta_k \hat{V}(\tilde{x}_t) \triangleq \sum_{g=1}^{k} \left(\tilde{x}_{t+g}^T \hat{P}_g(v_{t+g})\tilde{x}_{t+g} - \tilde{x}_t^T \hat{P}_g(v_t)\tilde{x}_t \right), \tag{4.22}$$

$$\Pi_L \triangleq \sum_{i_L=1}^{r} \sum_{j_L=1}^{r} \cdots \sum_{i_0=1}^{r} \sum_{j_0=1}^{r} \mu_{i_0}(v_t)\mu_{j_0}(v_t)\ldots\mu_{i_L}(v_{t+L})\mu_{j_L}(v_{t+L})\tilde{\Pi}_{L,i_0 j_0\ldots i_L j_L}, \tag{4.23}$$

$$\hat{\Pi}_L \triangleq G^{-T} \Pi_L G^{-1}. \tag{4.24}$$

As in the last theorem, we need to follow k steps to show that the conditions of the theorem satisfy H_∞ performance and stability of closed-loop system. There are some important differences. For example, in the gth step, we add and subtract $\tilde{x}_{t+k-g}^T \hat{\Pi}_{k-g} \tilde{x}_{t+k-g} + \frac{1}{\gamma} z_{t+k-g}^T z_{t+k-g} - \gamma w_{t+k-g}^T w_{t+k-g}$, and the nonlinear transformation (change of variables) will be as follows:

$$\begin{bmatrix} \mathcal{A}(v_t, v_t) & \mathcal{B}(v_t) \\ \mathcal{C}(v_t) & \mathcal{D}(v_t) \end{bmatrix} \triangleq \begin{bmatrix} M & YB_x(v_t) \\ 0 & I \end{bmatrix} \begin{bmatrix} \hat{A}(v_t, v_t) & \hat{B}(v_t) \\ \hat{C}(v_t) & \hat{D}(v_t) \end{bmatrix}$$
$$\times \begin{bmatrix} N & 0 \\ C_y(v_t)X & I \end{bmatrix} + \begin{bmatrix} YA(v_t)X & 0 \\ 0 & 0 \end{bmatrix}. \tag{4.25}$$

The rest of the proof is the same. $\square$

Remark 4.1. Theorem 4.1 involves the same number of LMIs as in Theorem 3.1 and $(n_p + n_c + n_\Pi + 3n^2 + 1)$ scalar variables, where $n_c = r\left(rn^2 + nm + np_y + mp_y\right)$.

Remark 4.2. In Theorem 4.1, if $\bar{P}^{ih}_{g,11}$, $\bar{P}^{ih}_{g,22}$, and $\bar{P}^{ih}_{g,21}$ for all $g = 2, \ldots, k$ are assumed to be zero, then this approach is reduced to the k-sample variation approach [16,15,12–14].

Remark 4.3. It is important to mention that the goal of using the theorem is to synthesize a controller with H_∞ performance when there is no solution using the standard Lyapunov theory. If we have no any feasible controller using the LMI tool-box in Matlab software, then we try for $k = 2$. If we cannot find any solution again, we will try for $k = 3$, and so on.

4.4 SIMULATION RESULTS

Consider the two-rule model of the unstable T-S fuzzy system with uncertainties in the form of (4.1), where

$$A_1 = \begin{bmatrix} -1.9 & -3.5 \\ 2 & 1 \end{bmatrix}, \quad A_2 = \begin{bmatrix} -0.08 & 0.8 \\ 0.58 & 0.69 \end{bmatrix}, \quad B_{x,1} = \begin{bmatrix} 0.26 \\ 0.92 \end{bmatrix},$$

$$B_{x,2} = \begin{bmatrix} -0.9 \\ 0.21 \end{bmatrix}, \quad D_{x,1} = D_{x,2} = \begin{bmatrix} 0 \\ 0.1 \end{bmatrix}, \quad C_{y,1} = C_{y,2} = \begin{bmatrix} 1 & 0 \end{bmatrix},$$

$$D_{y,1} = D_{y,2} = 0, \quad C_{z,1} = C_{z,2} = \begin{bmatrix} 0 & 1 \\ 1 & 0 \end{bmatrix}, \quad B_{z,1} = B_{z,2} = \begin{bmatrix} 0 \\ 0 \end{bmatrix}, \quad H = 0.1I_2,$$

$$E_1^a = E_2^a = \begin{bmatrix} 0.05 & 0 \\ 0 & 0.01 \end{bmatrix}, \quad E_1^b = \begin{bmatrix} 0.06 \\ 0 \end{bmatrix}, \quad E_2^b = \begin{bmatrix} -0.05 \\ 0 \end{bmatrix}.$$

The robust H_∞ output feedback control is designed in the following two cases:

Case 1: the approach proposed in Theorem 4.1.
Case 2: the k-sample variation approach [16,15,12–14].

The optimal H_∞ performance $\gamma_{\min}$ is computed for each of the aforementioned cases with $k = 1, 2$, and 3. The values of parameters ρ_1, ρ_2, and ρ_3 are assigned to be 0.1. It is worth noting that $k = 1$ corresponds to the standard Lyapunov theory. Table 4.1 shows the optimal $\gamma_{\min}$ for all cases with different values of k. We can observe that $\gamma_{\min}$ obtained using the developed method is smaller than $\gamma_{\min}$ obtained using the k-sample variation approach [16,15,12–14]. As expected, we can notice that $\gamma_{\min}$ decreases as k increases for both cases. Indeed, gaining smaller $\gamma_{\min}$ implies that results are less conservative, which clearly verifies our theoretical results. In addition, the numbers of LMIs and scalar variables in each case for different values of k are demonstrated in Table 4.1. The numbers of LMIs

TABLE 4.1 γ_{min}, number of LMIs and scalar variables in each case.

	k = 1	k = 2	k = 3
Theorem 4.1	solver time: 0.4840″ $\gamma_{min} = 5.8664$ 2 LMIs of 4 × 4 8 LMIs of 19 × 19 59 variables	$\gamma_{min} = 5.2244$, solver time: 2.4850″ 4 LMIs of 4 × 4 48 LMIs of 19 × 19 239 variables	$\gamma_{min} = 4.8902$, solver time: 15.1840″ 6 LMIs of 4 × 4 208 LMIs of 19 × 19 899 variables
k-samples variation [16,15,12–14]		$\gamma_{min} = 5.3725$, solver time: 2.2030″ 2 LMIs of 4 × 4 48 LMIs of 19 × 19 219 variables	$\gamma_{min} = 5.0034$, solver time: 14.9200″ 2 LMIs of 4 × 4 208 LMIs of 19 × 19 859 variables

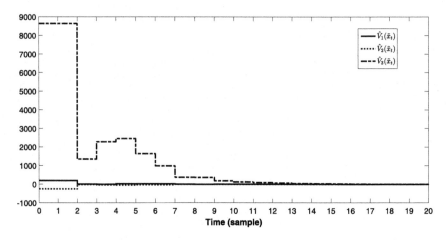

FIGURE 4.1 $\hat{V}_1(\tilde{x}_t)$, $\hat{V}_2(\tilde{x}_t)$, and $\hat{V}_3(\tilde{x}_t)$ in Theorem 4.1 with $k = 3$.

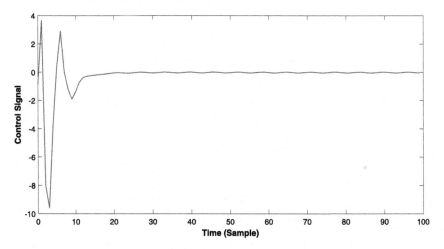

FIGURE 4.2 Control signal in simulation results using Theorem 4.1 when $k = 3$.

and scalar variables in the table confirm the formula in Remark 4.1, where $r = 2$, $n = 2$, $d = 1$, $m = 1$, $p = 2$, and $p_y = 1$. Now, to do simulation for closed-loop system, we set the initial conditions $x_t = (1, 1)$ and $\hat{x}_t = (0.5, 0)$ with zero disturbance. With the designed controller using Theorem 4.1 for $k = 3$, $\hat{V}_1(\tilde{x}_t)$, $\hat{V}_2(\tilde{x}_t)$, and $\hat{V}_3(\tilde{x}_t)$ are depicted in Fig. 4.1. From the figure we observe that $\hat{V}_3(\tilde{x}_t) > 0$ but $\hat{V}_1(\tilde{x}_t)$ and $\hat{V}_2(\tilde{x}_t)$ are not positive-definite. Moreover, in some samples, $\hat{V}_3(\tilde{x}_t)$ is increasing, but, finally, all of them converge to zero. We can see that the developed controller can effectively guarantee the stability of the resulting closed-loop systems. In addition, the control signal in Fig. 4.2 and output signal in Fig. 4.3 are shown for this case.

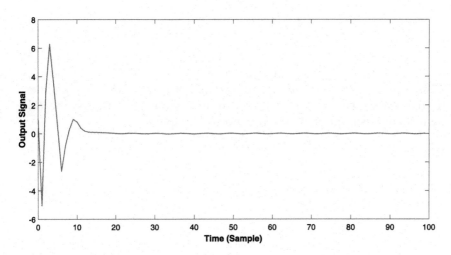

FIGURE 4.3 Output signal in simulation results using Theorem 4.1 when $k = 3$.

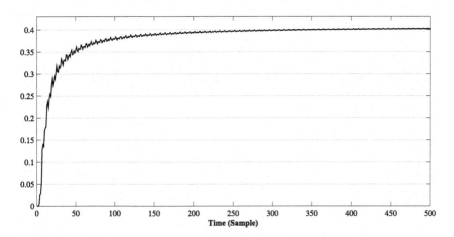

FIGURE 4.4 The ratio of the controlled output energy to the disturbance γ_{min} using Theorem 4.1 when $k = 3$.

The designed controller for $k = 3$ is associated with the system and efficiently stabilizes the system states. For simulations, $v_1(t) = x_1(t)$ is the premise variable, and $\mu_1(v_1(t)) = 0.5 + 0.5 \cos v_1(t)$ and $\mu_2(v_1(t)) = 1 - \mu_1(v_1(t))$ with sampling time 0.01. The obtained H_∞ performance is 0.4073, as shown in Fig. 4.4, which is the ratio of the controlled output z_t energy to the disturbance energy $w_t = 0.01 + 0.05 \sin t$. This implies that the L_2 gain from the disturbance to the regulated output is no greater than $\gamma_{min} = 0.4073$ and is much less than prescribed value of $\gamma_{min} = 4.8902$.

4.5 CONCLUSION

In this chapter, NLFs have been used to design robust output feedback H_∞ controllers for uncertain Takagi–Sugeno (T-S) fuzzy systems, and sufficient conditions have been given in terms of linear matrix inequalities (LMIs). It has been shown that the conservatism is reduced and the H_∞ performance is improved via a numerical example.

References

[1] T. Takagi, M. Sugeno, Fuzzy identification of systems and its applications to modeling and control, IEEE Transactions on Systems, Man and Cybernetics SMC-15 (1) (1985) 116–132.

[2] G. Feng, A survey on analysis and design of model-based fuzzy control systems, IEEE Transactions on Fuzzy Systems 14 (5) (2006) 676–697.

[3] W. Assawinchaichote, S.K. Nguang, P. Shi, E.-K. Boukas, H_∞ fuzzy state-feedback control design for nonlinear systems with stability constraints: an LMI approach, Mathematics and Computers in Simulation 78 (4) (2008) 514–531.

[4] J. Lam, S. Zhou, Dynamic output feedback H_∞ control of discrete-time fuzzy systems: a fuzzy-basis-dependent Lyapunov function approach, International Journal of Systems Science 38 (1) (2007) 25–37.

[5] S.K. Nguang, P. Shi, H_∞ fuzzy output feedback control design for nonlinear systems: an LMI approach, IEEE Transactions on Fuzzy Systems 11 (3) (2003) 331–340.

[6] M.H. Asemani, V.J. Majd, A robust H_∞ observer-based controller design for uncertain T-S fuzzy systems with unknown premise variables via LMI, Fuzzy Sets and Systems 2012 (2012) 21–40.

[7] E.F. Stephen Boyd, Laurent El Ghaoui, V. Balakrishnan, Linear Matrix Inequalities in System and Control Theory, Society for Industrial and Applied Mathematics (SIAM), 1994.

[8] G. Feng, Stability analysis of discrete-time fuzzy dynamic systems based on piecewise Lyapunov functions, IEEE Transactions on Fuzzy Systems 12 (1) (2004) 22–28.

[9] H.-N. Wu, K.-Y. Cai, H2 guaranteed cost fuzzy control design for discrete-time nonlinear systems with parameter uncertainty, Automatica 42 (7) (2006) 1183–1188.

[10] S.H. Kim, Improved approach to robust $\mathcal{H}_\infty$ stabilization of discrete-time T-S fuzzy systems with time-varying delays, IEEE Transactions on Fuzzy Systems 18 (5) (2010) 1008–1015.

[11] A. Ahmadi, P. Parrilo, Non-monotonic Lyapunov functions for stability of discrete time nonlinear and switched systems, in: 47th IEEE Conference on Decision and Control, 2008 (CDC 2008), 2008, pp. 614–621.

[12] T. Guerra, A. Kruszewski, M. Bernal, Control law proposition for the stabilization of discrete Takagi–Sugeno models, IEEE Transactions on Fuzzy Systems 17 (3) (2009) 724–731.

[13] A. Kruszewski, R. Wang, T. Guerra, Nonquadratic stabilization conditions for a class of uncertain nonlinear discrete time T-S fuzzy models: a new approach, IEEE Transactions on Automatic Control 53 (2) (2008) 606–611.

[14] Y.-J. Chen, M. Tanaka, K. Inoue, H. Ohtake, K. Tanaka, T. Guerra, A. Kruszewski, H. Wang, A nonmonotonically decreasing relaxation approach of Lyapunov functions to guaranteed cost control for discrete fuzzy systems, IET Control Theory & Applications 8 (16) (2014) 1716–1722.

[15] S. Derakhshan, A. Fatehi, M. Sharabiany, Nonmonotonic observer-based fuzzy controller designs for discrete time T-S fuzzy systems via LMI, IEEE Transactions on Cybernetics 44 (12) (2014) 2557–2567.

[16] S.F. Derakhshan, A. Fatehi, Non-monotonic robust H2 fuzzy observer-based control for discrete time nonlinear systems with parametric uncertainties, International Journal of Systems Science 46 (12) (2015) 2134–2149.

[17] S.F. Derakhshan, A. Fatehi, Non-monotonic Lyapunov functions for stability analysis and stabilization of discrete time Takagi–Sugeno fuzzy systems, International Journal of Innovative Computing, Information & Control (2014) 1567–1586.

[18] E. Kim, H. Lee, New approaches to relaxed quadratic stability condition of fuzzy control systems, IEEE Transactions on Fuzzy Systems 8 (5) (2000) 523–534.

[19] W.-J. Wang, Y.-J. Chen, C.-H. Sun, Relaxed stabilization criteria for discrete-time T-S fuzzy control systems based on a switching fuzzy model and piecewise Lyapunov function, IEEE Transactions on Systems, Man and Cybernetics. Part B. Cybernetics 37 (3) (2007) 551–559.

5

Stability and H_∞ Control of Discrete-Time Switched Systems via One-Step Ahead Lyapunov Function Approach

5.1 INTRODUCTION

Switched systems have drawn an everincreasing attention in recent years motivated by either technological or economical reasons. They play essential roles in numerous physical and engineering problems whenever the switching properties of a dynamical system have to be considered. For example, Markovian jump systems (MJSs), which are governed by a Markov process with complex and time-varying transition probabilities, have exerted a profound and widespread influence on the random system theory [1,2]. Switched systems with slowly switching rate start to penetrate into the research frontier due to their high degree of design freedom [3–6]. The most general case is the so-called arbitrarily switched systems, which exhibit arbitrary switching among a family of subsystems depending on various environmental or inherent factors. Some related topics, such as adaptive neural output-feedback control [7] or adaptive tracking control [8] for nonlinear switched systems, H_∞ filtering problem for nonlinear switched systems with both stable and unstable subsystems [9], sampled data stabilization [10], and finite-time H_∞ control [11] for switched systems under dwell-time constraints, have also received burgeoning research interests.

On the constructive methods for stability and stabilization of arbitrarily switched systems, it is well known that the globally uniform asymptotical stability (GUAS) property is a consequence of the existence of a common Lyapunov function (LF) [12–14]. The concept of a common LF is useful from a practical perspective because for a certain class of the underlying systems, if a common LF exists for all subsystems, then it is always possible to find an LF of certain type such as, quadratic, piecewise quadratic, bounded polynomial, etc. However, the stability criterion developed by a common LF is very conservative because searching for a unique Lyapunov matrix for a set of models may become difficult, even infeasible, especially when the number of submodels greatly increases. Although arbitrary switching mechanism has already been included in many general dynamical models such as stochastic nonlinear systems [15], neural-networked control systems [16], nonlinear triangular systems [17], etc., we believe that the fundamental theory, such as GUAS property, still has great potential to be improved.

On another research frontline, the nonmonotonic LF approach, which is developed for T-S fuzzy model to relax the monotonicity requirement of fuzzy LFs, has attracted burgeoning research interests in recent years. Several effective conservatism reduction techniques, such as LF of two-samples variations $V_{k+2} - V_k < 0$ [18–20], LF of N-sample variations $V_{k+N} - V_k < 0$ [21–26], and the most general case $V_{k+N}^N - V_k^N + \cdots + V_{k+2}^2 - V_k^2 + V_{k+1}^1 - V_k^1 < 0$ [27], have been intensively studied.

Motivated by these observations and since the T-S fuzzy model can be viewed as a particular case of the general switching, that is, soft switching according to the membership function, we aim at formulating a one-step ahead LF that can ensure GUAS property for arbitrarily switched systems with further less conservativeness. There are two main contributions. First, the concept of one-step ahead LF is introduced by incorporating the future state one-step ahead of the current time. The LF matrix has a more general structure than the existing nonmonotonic LF, i.e., $\mathcal{L}(x_k) = V^1(x_k) + V^2(x_k) + V^2(x_{k+1})$. Second, a mode-independent H_∞ controller is designed by eliminating the product term between the LF matrix and augmented system matrix. The overall procedure is proceeded without any structural constraint on the $\mathcal{P}$ matrix.

Notations: Notations used in this chapter are standard. $\mathbb{R}^n$ and $\mathbb{R}^{n \times m}$ denote the n-dimensional Euclidean space and the set of all $n \times m$ matrices, respectively. By $\mathbb{N}$ we denote the set of positive integers, and $\| \cdot \|_2$ refers to the Euclidean vector norm. The brief notation $P > 0 (\geq 0)$ means that a matrix P is real symmetric and positive definite (semipositive definite). The zero matrix of size $n \times m$ is denoted by 0_{nm}, and I_n represents the identity matrix of size $n \times n$; diag$\{\cdot\}$ and exp$\{\cdot\}$ represent the diagonal matrix and matrix exponential function, respectively; $l_2[0, M]$ is the space of summable sequences on $[0, M]$, where M may be finite or infinite. A function $\kappa : [0, \infty) \to [0, \infty)$ is said to belong to class $\mathcal{K}_\infty$ if it is continuous, strictly increasing, and radially unbounded and satisfies $\kappa(0) = 0$. Similarly, a function $\varrho : [0, \infty) \times [0, \infty) \to [0, \infty)$ is said to belong to class $\mathcal{KL}$ if, for each fixed t, the function $\varrho(r, t)$ belongs to class $\mathcal{K}_\infty$ and, for each fixed r, the function $\varrho(r, t)$ decreases to 0 as $t \to \infty$.

5.2 PRELIMINARIES AND PROBLEM FORMULATION

Consider a class of linear discrete-time systems with arbitrarily switching dynamics

$$\begin{cases} x_{k+1} = A(\sigma_k)x_k + B(\sigma_k)u_k + F(\sigma_k)w_k, \\ z_k = C(\sigma_k)x_k + D(\sigma_k)u_k + H(\sigma_k)w_k, \end{cases} \tag{5.1}$$

where $k \in \{0, 1, \cdots, M\}$, $x_k \in \mathbb{R}^{n_x}$ is the state vector, $u_k \in \mathbb{R}^{n_u}$ is the control input, $z_k \in \mathbb{R}^{n_z}$ is the controlled output, and $w_k \in \mathbb{R}^{n_w}$ is the exogenous disturbance satisfying $\{w_k\} \in l_2[0, M]$. When $w = 0$ and $u = 0$, system (5.1) can be viewed as a *free* switched system. When $u = 0$, the underlying system refers to as an *input-free* switched system. In the case $w = 0$, it is usually named a *disturbance-free* switched system.

The switching signal σ is a piecewise constant function of time taking values in a finite integer set $\mathbb{S} = \{1, 2, \cdots, s\}$, where $s \in \mathbb{N}$ represents the number of subsystems; σ may depend on any deterministic, logic, or stochastic rules, and it is assumed to be right-continuous. As commonly assumed in the literature, the Zeno behavior and jumps of the state at the switching instants are not considered in this work. For each possible value of $\sigma_k = i_0$, we denote by $\mathcal{N}_{i_0}$ the matrices associated with the "i_0th subsystem," where $\mathcal{N}$ can be replaced by any symbols. For example, A_{i_0}, B_{i_0}, F_{i_0}, C_{i_0}, D_{i_0}, and H_{i_0} are used to denote constant matrices depending on the switching signal σ_k. Similarly, we can define the cases where $\sigma_{k+1} = i_1$, $\sigma_{k+2} = i_2, \ldots$, etc. The brief notations of subsystems satisfy $i_0, i_1, i_2 \in \mathbb{S}$.

The mode-independent state feedback controller is of the form

$$u_k = Kx_k = YG^{-1}x_k. \tag{5.2}$$

Thus, the closed-loop system can be written as

$$\begin{cases} x_{k+1} = \tilde{A}_{i_0}x_k + F_{i_0}w_k, \\ z_k = \tilde{C}_{i_0}x_k + H_{i_0}w_k, \end{cases} \tag{5.3}$$

where $\tilde{A}_{i_0} = A_{i_0} + B_{i_0}K$ and $\tilde{C}_{i_0} = C_{i_0} + D_{i_0}K$.

By denoting four new vectors $\zeta_k \triangleq [x_k^T \ x_{k+1}^T]^T$, $\xi_k \triangleq [z_k^T \ z_{k+1}^T]^T$, $\upsilon_k \triangleq [u_k^T \ u_{k+1}^T]^T$, and $\varpi_k \triangleq [w_k^T \ w_{k+1}^T]^T$ we have the following augmented system:

$$\begin{cases} \zeta_{k+1} = \mathcal{A}_{i_0,i_1}\zeta_k + \mathcal{B}_{i_0,i_1}\upsilon_k + \mathcal{F}_{i_0,i_1}\varpi_k, \\ \xi_k = \mathcal{C}_{i_0,i_1}\zeta_k + \mathcal{D}_{i_0,i_1}\upsilon_k + \mathcal{H}_{i_0,i_1}\varpi_k, \end{cases} \tag{5.4}$$

where the matrix $\mathcal{M}_{i_0,i_1}$ noted in (5.4) is represented by $\mathcal{M}_{i_0,i_1} = \text{diag}\{\mathcal{M}_{i_0}, \mathcal{M}_{i_1}\}$.

Subsequently, the closed-loop form of (5.4) can be further written as

$$\begin{cases} \zeta_{k+1} = \tilde{A}_{i_0,i_1} \zeta_k + \mathcal{F}_{i_0,i_1} \varpi_k, \\ \xi_k = \tilde{C}_{i_0,i_1} \zeta_k + \mathcal{H}_{i_0,i_1} \varpi_k, \end{cases} \tag{5.5}$$

where

$$\tilde{A}_{i_0,i_1} = \text{diag}\left\{ A_{i_0} + B_{i_0} K, A_{i_1} + B_{i_1} K \right\},$$

$$\tilde{C}_{i_0,i_1} = \text{diag}\left\{ C_{i_0} + D_{i_0} K, C_{i_1} + D_{i_1} K \right\}.$$

The purpose of this chapter is to design a state feedback controller of the form (5.2) such that the following two conditions are satisfied:
i) The closed-loop system (5.3) is GUAS.
ii) Under the zero-initial condition, the controlled output z satisfies

$$\|z\|_2^2 < \gamma^2 \|w\|_2^2 \quad \text{or} \quad \|\xi\|_2^2 < \gamma^2 \|\varpi\|_2^2, \tag{5.6}$$

where

$$\|z\|_2^2 = \sum_{k=0}^{M} z_k^\mathrm{T} z_k < \infty \quad \text{or} \quad \|\xi\|_2^2 = \sum_{k=-1}^{M-1} \xi_k^\mathrm{T} \xi_k < \infty \tag{5.7}$$

for any nonzero $w_k \in l_2[0, M]$ and an optimized attenuation level $\gamma > 0$.
Before proceeding further, we recall the following preliminaries.

Definition 5.1. The *free* switched system (5.1) is said to be GUAS if there exists a function ϱ of class $\mathcal{KL}$ such that for arbitrarily switching signal σ and all initial conditions x_{k_0}, the solutions of dynamic model (5.1) satisfy the inequality $\|x_k\| \le \varrho(\|x_0\|, k)$ for all $k \ge k_0$.

Definition 5.2. The LF $\mathcal{L}(\zeta_k)$ is said to be a one-step ahead LF of switched system (5.1) with any switching signal σ if it satisfies

$$\mathcal{L}(\zeta_k) \triangleq \zeta_k^\mathrm{T} \mathcal{P} \zeta_k, \tag{5.8}$$

where

$$\zeta_k \triangleq [x_k^\mathrm{T} \ x_{k+1}^\mathrm{T}]^\mathrm{T}, \quad \mathcal{P} = \begin{bmatrix} P_{11} & P_{12} \\ P_{21} & P_{22} \end{bmatrix} > 0.$$

Lemma 5.1. *[27] The free switched system (5.1) is GUAS if there exist matrices P_1 and P_2 such that*

$$P_2 > 0, \ P_1 + P_2 > 0 \tag{5.9a}$$

$$- P_1 - P_2 + A_{i_0}^\mathrm{T} P_1 A_{i_0} + A_{i_0}^\mathrm{T} A_{i_1}^\mathrm{T} P_2 A_{i_1} A_{i_0} < 0. \tag{5.9b}$$

Proof. Consider a Lyapunov-like function

$$\mathcal{L}(x_k) \triangleq V^1(x_k) + V^2(x_k) + V^2(x_{k+1}), \tag{5.10}$$

where $V^1(x_k) = x_k^T P_1 x_k$ and $V^2(x_k) = x_k^T P_2 x_k$. We can see from (5.9a) that $V^1(x_k)$ is not a Lyapunov-like function. Then we have

$$\Delta\mathcal{L}(x_k) = V^2(x_{k+2}) - V^2(x_k) + V^1(x_{k+1}) - V^1(x_k). \tag{5.11}$$

Now we can prove the lemma following the lines of the proof of Theorem 2.2 in [27]. □

5.3 STABILITY AND H_∞ PERFORMANCE

In this section, we first present a relaxed sufficient condition to achieve the GUAS property of *free* switched (5.1) by constructing a one-step ahead LF. Since exogenous disturbances ubiquitously exist in practical systems, we further address some results on the analysis of H_∞ performance.

Theorem 5.1. *The* free *switched system (5.1) is GUAS if there exist matrices* $\mathcal{P} > 0$ *and* $\mathcal{Q}_{i_0,i_1}$ *such that*

$$-\mathcal{P} - \mathcal{Q}_{i_0,i_1} + \mathcal{A}_{i_0,i_1}^T \mathcal{P} \mathcal{A}_{i_0,i_1} < 0, \tag{5.12a}$$

$$\Theta_{i_0,i_1} = \begin{bmatrix} I_{nx} & A_{i_0}^T \end{bmatrix} \mathcal{Q}_{i_0,i_1} \begin{bmatrix} I_{nx} \\ A_{i_0} \end{bmatrix} \leq 0. \tag{5.12b}$$

Proof. Consider the one-step ahead LF (5.8). The condition $\mathcal{P} > 0$ implies that

$$\begin{cases} \mathcal{L}(\zeta) \text{ are radially unbounded,} \\ \mathcal{L}(\zeta) > 0 \,\forall \zeta \neq 0, \\ \mathcal{L}(0) = 0. \end{cases} \tag{5.13}$$

Condition (5.12a) gives that

$$\Delta\mathcal{L}(\zeta_k) = \mathcal{L}(\zeta_{k+1}) - \mathcal{L}(\zeta_k) = \zeta_k^T \left\{ \mathcal{A}_{i_0,i_1}^T \mathcal{P} \mathcal{A}_{i_0,i_1} - \mathcal{P} \right\} \zeta_k < \zeta_k^T \mathcal{Q}_{i_0,i_1} \zeta_k. \tag{5.14}$$

Condition (5.12b) results in

$$\zeta_k^T \mathcal{Q}_{i_0,i_1} \zeta_k = x_k^T \Theta_{i_0,i_1} x_k \leq 0. \tag{5.15}$$

Obviously, inequalities (5.14) and (5.15) guarantee that $\Delta\mathcal{L}(\zeta_k) < 0$, i.e.,

$$\Delta\mathcal{L}(\zeta_k) < -\beta_{i_0,i_1} x_k^T x_k \leq -\beta x_k^T x_k, \tag{5.16}$$

where $\beta_{i_0,i_1} = \lambda_{\min}(-\Theta_{i_0,i_1})$ and $\beta = \min_{i_0,i_1 \in \mathbb{S}} \beta_{i_0,i_1}$.

Therefore, we have that, for any $M \geq 1$,

$$\sum_{k=0}^{M} \Delta \mathcal{L}(\zeta_k) = \mathcal{L}(\zeta_{M+1}) - \mathcal{L}(\zeta_0) < -\beta \sum_{k=0}^{M} \|x_k\|_2^2. \tag{5.17}$$

This yields that

$$\sum_{k=0}^{M} \|x_k\|_2^2 < \beta^{-1} \left(\mathcal{L}(\zeta_0) - \mathcal{L}(\zeta_{M+1}) \right) \leq \beta^{-1} \mathcal{L}(\zeta_0), \tag{5.18}$$

which implies

$$\lim_{M \to \infty} \left\{ \sum_{k=0}^{M} \|x_k\|_2^2 \right\} < \beta^{-1} \mathcal{L}(\zeta_0) < \infty. \tag{5.19}$$

Therefore, we can conclude from (5.13), (5.19), and Definition 5.1 that the *free* switched system (5.1) is GUAS. $\qquad\qquad\square$

Remark 5.1. Note that there is no negative constraint on the matrices $\mathcal{Q}_{i_0,i_1}$ and they can be viewed as slack variables to further reduce the conservatism of the stability criterion. The matrices $\mathcal{P}$ and $\mathcal{Q}_{i_0,i_1}$ in Theorem 5.1 have no any structural constraints.

Remark 5.2. We can prove that the one-step ahead approach claimed in Theorem 5.1 covers the nonmonotonic method considering two-sample variations, i.e., $V_{k+2}^2 - V_k^2 + V_{k+1}^1 - V_k^1 < 0$ [27], or $V_{k+2} - V_k < 0$ [18–20], or the S-variable approach [28,29] as a particular case.

• In Lemma 5.1, if we let $A_{i1}^{\mathrm{T}} P_2 A_{i1} < Q$ and keep in mind that there is no constraint on the matrix Q, then condition (5.9b) can be relaxed as

$$\begin{cases} -Q + A_{i1}^{\mathrm{T}} P_2 A_{i1} < 0, \\ -P_1 - P_2 + A_{i0}^{\mathrm{T}} (P_1 + Q) A_{i0} < 0. \end{cases} \tag{5.20}$$

Initially, let us introduce the special matrices

$$\mathcal{P} = \mathrm{diag}\{P_1 + P_2, P_2\}, \tag{5.21}$$

$$\mathcal{Q}_{i_0,i_1} = \mathrm{diag}\left\{-A_{i_0}^{\mathrm{T}}(Q - P_2)A_{i_0}, -P_2 + Q\right\} \tag{5.22}$$

in Theorem 5.1. Then we can directly conclude that $\mathcal{P} > 0$ covers condition (5.9a).

Furthermore, condition (5.12a) can be rewritten as

$$\mathrm{diag}\left\{-P_1 - P_2 + A_{i_0}^{\mathrm{T}}(P_1 + Q)A_{i_0}, -Q + A_{i_1}^{\mathrm{T}} P_2 A_{i_1}\right\} < 0$$

with some simple mathematical arrangements, which directly implies (5.20) and then covers (5.9b) in Lemma 5.1.

Additionally, if we substitute (5.22) into Θ_{i_0,i_1} in (5.12b), then we obtain $\Theta_{i_0,i_1} = 0$.

Therefore, we can conclude that the one-step ahead LF approach based on (5.12) covers the nonmonotonic LF method based on (5.9), i.e., $V_{k+2}^2 - V_k^2 + V_{k+1}^1 - V_k^1 < 0$, as a particular case.

• Intuitively, $V_{k+2}^2 - V_k^2 + V_{k+1}^1 - V_k^1 < 0$ implies $V_{k+2}^2 - V_k^2 < 0$ or $V_{k+1}^1 - V_k^1 < 0$, and therefore it covers $V_{k+2} - V_k < 0$ as a particular case. In other words, $V_{k+2} - V_k < 0$ is also a particular case of the one-step ahead LF approach presented in Theorem 5.1.

• It is worth mentioning that the S-variable approach can also be extended to switched systems. However, the introduced auxiliary variables cannot improve the stability analysis, and it is an equivalence of the common Lyapunov inequality, which implies $V_{k+1} - V_k < 0$. The main advantage lies in dealing with the structural uncertainties.

Remark 5.3. It is clear that Theorem 5.1 is easy to be extended to the multistep ahead case, i.e., N-step ahead of the current time, only by piling up $A_{i_{N-1}}, \ldots, A_{i_1}, A_{i_0}$. Therefore, it also has great potential to cover the results of $V_{k+N} - V_k < 0$ [21–26] and $V_{k+N}^N - V_k^N \cdots + V_{k+2}^2 - V_k^2 + V_{k+1}^1 - V_k^1 < 0$ [27]. Note that we usually will not go further to N-step ahead due to its inherent complexity in the controller, robustness, and H_∞ synthesis, and therefore we give the one-step ahead case in this study for the briefness of the proof.

Based on the results of GUAS property, we further pay our attention to the H_∞ performance analysis.

Theorem 5.2. *The input-free switched system (5.1) is GUAS and keeps a disturbance attenuation level γ for all $\varpi \in l_2[0, M]$, $\varpi \neq 0$, if there exist matrices $\mathcal{P} > 0$ and $\mathcal{Q}_{i_0,i_1}$ such that*

$$\Xi_{i_0,i_1} = \begin{bmatrix} \Xi_{i_0,i_1}^{11} & \Xi_{i_0,i_1}^{12} \\ * & \Xi_{i_0,i_1}^{22} \end{bmatrix} < 0, \tag{5.23a}$$

$$\begin{bmatrix} I_{nx} & A_{i_0}^{\mathrm{T}} \\ 0_{n_w,n_x} & F_{i_0}^{\mathrm{T}} \end{bmatrix} \mathcal{Q}_{i_0,i_1} \begin{bmatrix} I_{nx} & 0_{n_x,n_w} \\ A_{i_0} & F_{i_0} \end{bmatrix} \leq 0, \tag{5.23b}$$

where

$$\Xi_{i_0,i_1}^{11} = -\mathcal{P} - \mathcal{Q}_{i_0,i_1} + \mathcal{A}_{i_0,i_1}^{\mathrm{T}} \mathcal{P} \mathcal{A}_{i_0,i_1} + \mathcal{C}_{i_0,i_1}^{\mathrm{T}} \mathcal{C}_{i_0,i_1},$$

$$\Xi_{i_0,i_1}^{12} = \mathcal{A}_{i_0,i_1}^{\mathrm{T}} \mathcal{P} \mathcal{F}_{i_0,i_1} + \mathcal{C}_{i_0,i_1}^{\mathrm{T}} \mathcal{H}_{i_0,i_1},$$

$$\Xi_{i_0,i_1}^{12} = -\gamma^2 I_{2n_w} + \mathcal{F}_{i_0,i_1}^{\mathrm{T}} \mathcal{P} \mathcal{F}_{i_0,i_1} + \mathcal{H}_{i_0,i_1}^{\mathrm{T}} \mathcal{H}_{i_0,i_1}.$$

Proof. Obviously, condition (5.23) implies the inequalities in (5.12), and hence GUAS property of the free system is proved.

Assume that the input-free system (5.1) has a zero initial condition, that is, $x_k = 0, k \in \{-1, 0\}$, and define

$$\varsigma_k \triangleq \left[\zeta_k^{\mathrm{T}}, \varpi_k^{\mathrm{T}} \right]^{\mathrm{T}}, \quad \Xi'_{i_0,i_1} \triangleq \Xi_{i_0,i_1} + \left[\begin{array}{c|c} \mathcal{Q}_{i_0,i_1} & 0_{2n_x,2n_w} \\ \hline * & 0_{2n_w,2n_w} \end{array} \right],$$

$$J_M \triangleq \|\xi\|_2^2 - \gamma^2 \|\varpi\|_2^2 = \sum_{k=-1}^{M-1} \left\{ \xi_k^{\mathrm{T}} \xi_k - \gamma^2 \varpi_k^{\mathrm{T}} \varpi_k \right\}.$$

Since $\mathcal{L}(\varsigma_{-1}) = 0$, $\mathcal{L}(\varsigma_0) \neq 0$, and $\mathcal{L}(\varsigma_M) \geq 0$, we get

$$\begin{aligned}
J_M &= \sum_{k=-1}^{M-1} \left\{ \Delta\mathcal{L}(\varsigma_k) + \xi_k^{\mathrm{T}} \xi_k - \gamma^2 \varpi_k^{\mathrm{T}} \varpi_k \right\} - \mathcal{L}(\varsigma_M) \\
&\leq \sum_{k=-1}^{M-1} \left\{ \Delta\mathcal{L}(\varsigma_k) + \xi_k^{\mathrm{T}} \xi_k - \gamma^2 \varpi_k^{\mathrm{T}} \varpi_k \right\} \qquad (5.24) \\
&= \sum_{k=-1}^{M-1} \left\{ \varsigma_k^{\mathrm{T}} \Xi'_{i_0,i_1} \varsigma_k \right\}.
\end{aligned}$$

Noting that condition (5.23) means $\varsigma_k^{\mathrm{T}} \Xi'_{i_0,i_1} \varsigma_k < 0$, we then have $J_M < 0$. Therefore, the dissipativity (5.6) holds for any finite or infinite number $M > 0$. This completes the proof. $\qquad \square$

Remark 5.4. The H_∞ analysis approach can be extended to an N-step ahead scenario by moving the horizon from $k + N - 1$ to $k + 1$ step by step. Due to the multistep feature, sufficient conditions can be expressed by a set of LMI constraints rather than by a big matrix inequality (see [30] for reference).

Remark 5.5. The one-step ahead LF approach has great potentials to be extended to multimodel systems, such as T-S fuzzy models, inhomogeneous Markovian jump systems (MJLSs), constrained switched systems, etc. However, it is worth mentioning that such an approach cannot be directly employed to MJLSs with time-invariant transition probabilities (TPs) because a necessary and sufficient condition has already been given by fully making use of the TPs knowledge. Moreover, when it is employed to the average dwell-time (ADT) switching systems, the essential difficulty is to construct an exponential damping law of the decreasing points of LF, and the ADT constraint should be further relaxed.

5.4 H_∞ CONTROLLER DESIGN

In this section, we consider the H_∞ disturbance attenuation for arbitrarily switched system (5.1) by using the one-step ahead LF approach to design a controller of the form (5.2).

Theorem 5.3. *For the switched system (5.1) with arbitrary switching signal σ, there exists a mode-independent controller (5.2) such that the closed-loop system is GUAS and possesses an attenuation level γ if there exist matrices $P_{11} > 0$, $P_{22} > 0$, $R > 0$, P_{12}, P_{21}, $Q^1_{i_0,i_1}$, $Q^2_{i_0,i_1}$, and G and a scalar $\alpha > 0$ such that*

$$
\Upsilon_{i_0,i_1} =
\begin{bmatrix}
\Upsilon^{11}_{i_0,i_1} & \Upsilon^{12}_{i_0,i_1} & \Upsilon^{13}_{i_0,i_1} & \Upsilon^{14}_{i_0,i_1} \\
* & \Upsilon^{22}_{i_0,i_1} & \Upsilon^{23}_{i_0,i_1} & 0_{2n_w,2n_z} \\
* & * & \Upsilon^{33}_{i_0,i_1} & 0_{2n_x,2n_z} \\
* & * & * & -I_{2n_z}
\end{bmatrix} < 0,
\tag{5.25a}
$$

$$
\Lambda_{i_0,i_1} =
\begin{bmatrix}
-R & 0_{n_x,n_w} & G^T & \Lambda^{14}_{i_0} \\
* & -\alpha I_{n_w} & 0_{n_w,n_x} & F^T_{i_0} \\
* & * & \Lambda^{33}_{i_0,i_1} & 0_{n_x,n_x} \\
* & * & * & \Lambda^{44}_{i_0,i_1}
\end{bmatrix} < 0,
\tag{5.25b}
$$

where

$$
\Upsilon^{11}_{i_0,i_1} =
\begin{bmatrix}
-P_{11} - Q^1_{i_0,i_1} + R & -P_{12} \\
-P_{21} & -P_{22} - Q^2_{i_0,i_1}
\end{bmatrix},
$$

$$
\Upsilon^{12}_{i_0,i_1} =
\begin{bmatrix}
\left(\begin{matrix} G^T C^T_{i_0} H_{i_0} \\ +Y^T D^T_{i_0} H_{i_0} \end{matrix} \right) & 0_{n_x,n_w} \\
\hline
0_{n_x,n_w} & \left(\begin{matrix} G^T C^T_{i_1} H_{i_1} \\ +Y^T D^T_{i_1} H_{i_1} \end{matrix} \right)
\end{bmatrix},
$$

$$
\Upsilon^{13}_{i_0,i_1} =
\begin{bmatrix}
G^T A^T_{i_0} + Y^T B^T_{i_0} & 0_{n_x,n_x} \\
0_{n_x,n_x} & G^T A^T_{i_1} + Y^T B^T_{i_1}
\end{bmatrix},
$$

$$
\Upsilon^{14}_{i_0,i_1} =
\begin{bmatrix}
G^T C^T_{i_0} + Y^T D^T_{i_0} & 0_{n_x,n_x} \\
0_{n_x,n_z} & G^T C^T_{i_1} + Y^T D^T_{i_1}
\end{bmatrix},
$$

$$\Upsilon_{i_0,i_1}^{22} = \left[\begin{array}{c:c} \begin{pmatrix} (-\gamma^2 + \alpha)I_{n_w} \\ + H_{i_0}^{\mathrm{T}} H_{i_0} \end{pmatrix} & 0_{n_w,n_w} \\ \hdashline 0_{n_w,n_w} & \begin{pmatrix} -\gamma^2 I_{n_w} \\ + H_{i_1}^{\mathrm{T}} H_{i_1} \end{pmatrix} \end{array} \right],$$

$$\Upsilon_{i_0,i_1}^{23} = \left[\begin{array}{c:c} F_{i_0}^{\mathrm{T}} & 0_{n_w,n_x} \\ \hdashline 0_{n_w,n_x} & F_{i_1}^{\mathrm{T}} \end{array} \right],$$

$$\Upsilon^{33} = \left[\begin{array}{cc} -G^{\mathrm{T}} - G + P_{11} & P_{12} \\ P_{21} & -G^{\mathrm{T}} - G + P_{22} \end{array} \right], \Lambda_{i_0}^{14} = G^{\mathrm{T}} A_{i_0} + Y^{\mathrm{T}} B_{i_0},$$

$$\Lambda_{i_0,i_1}^{33} = -G^{\mathrm{T}} - G + Q_{i_0,i_1}^1, \Lambda_{i_0,i_1}^{44} = -G^{\mathrm{T}} - G + Q_{i_0,i_1}^2.$$

Proof. Consider the closed-loop system (5.5) and one-step ahead LF as $\mathcal{L}(\zeta_k) \triangleq \zeta_k^{\mathrm{T}} \mathcal{G}^{-\mathrm{T}} \mathcal{P} \mathcal{G}^{-1} \zeta_k$, where

$$\mathcal{P} = \left[\begin{array}{cc} P_{11} & P_{12} \\ P_{21} & P_{22} \end{array} \right] > 0, \ \mathcal{G} = \mathrm{diag}\{G, G\}.$$

Denote the matrix in this theorem as

$$\mathcal{Q}_{i_0,i_1} = \mathrm{diag}\{Q_{i_0,i_1}^1, Q_{i_0,i_1}^2\}.$$

According to the proof of Theorem 5.2, we know $\Delta L(\zeta_k) + \xi_k^{\mathrm{T}} \xi_k - \gamma^2 \varpi_k^{\mathrm{T}} \varpi_k < 0$, which implies the GUAS property and H_∞ disturbance attenuation capability of the close-loop system (5.3). Let

$$\Delta L(\zeta_k) + \xi_k^{\mathrm{T}} \xi_k - \gamma^2 \varpi_k^{\mathrm{T}} \varpi_k < \xi_k^{\mathrm{T}} \mathcal{G}^{-\mathrm{T}} \mathcal{Q}_{i_0,i_1} \mathcal{G}^{-1} \xi_k$$
$$- x_k^{\mathrm{T}} G^{-\mathrm{T}} R G^{-1} x_k - \alpha w_k^{\mathrm{T}} w_k \quad (5.26)$$

and

$$\xi_k^{\mathrm{T}} \mathcal{G}^{-\mathrm{T}} \mathcal{Q}_{i_0,i_1} \mathcal{G}^{-1} \xi_k - x_k^{\mathrm{T}} G^{-T} R G^{-1} x_k - \alpha w_k^{\mathrm{T}} w_k < 0. \quad (5.27)$$

We know that (5.26) and (5.27) can guarantee the GUAS property and H_∞ performance. Note that (5.27) can be rewritten as

$$- \begin{bmatrix} x_k^{\mathrm{T}} & w_k^{\mathrm{T}} \end{bmatrix} \left[\begin{array}{c:c} -G^{-\mathrm{T}} R G^{-1} & 0_{n_x,n_w} \\ \hdashline 0_{n_w,n_x} & -\alpha I_{n_w} \end{array} \right] \begin{bmatrix} x_k \\ w_k \end{bmatrix}$$

$$+ \begin{bmatrix} x_k^{\mathrm{T}} & w_k^{\mathrm{T}} \end{bmatrix} \left[\begin{array}{c:c} I_{nx} & A_{i_0}^{\mathrm{T}} + K^{\mathrm{T}} B_{i_0}^{\mathrm{T}} \\ \hdashline 0_{n_w n_x} & F_{i_0}^{\mathrm{T}} \end{array} \right] \mathcal{G}^{-\mathrm{T}} \mathcal{Q}_{i_0,i_1} \mathcal{G}^{-1}$$

$$\times \left[\begin{array}{c:c} I_{nx} & 0_{n_x n_w} \\ \hdashline A_{i_0} + B_{i_0} K & F_{i_0} \end{array} \right] \begin{bmatrix} x_k \\ w_k \end{bmatrix} < 0. \quad (5.28)$$

Substituting (5.28) into (5.26) and (5.27) and applying some simple mathematical techniques, such as Schur complement and congruence transformation, we can obtain LMI constraints (5.25a) and (5.25b). This completes the proof. □

5.5 ILLUSTRATIVE EXAMPLES

In this section, we provide illustrative examples to verify the analysis, design procedure, and effectiveness of the proposed one-step ahead LF method.

Example 5.1. Consider an input-free switched system (5.1) with two subsystems, which is the discretization of the continuous time model given in [31]:

$$A_1 = \exp\left\{\begin{bmatrix} -0.1 & -0.1 \\ 0.1 & -0.1 \end{bmatrix}\right\}, F_1 = \begin{bmatrix} 0.4 \\ 0.5 \end{bmatrix}, C_1 = \begin{bmatrix} 0.2 & 0.1 \end{bmatrix}, H_1 = 0.1,$$

$$A_2 = \exp\left\{\begin{bmatrix} -0.1 & -0.1b \\ 0.1a & -0.1 \end{bmatrix}\right\}, F_2 = \begin{bmatrix} 0.2 \\ 0.6 \end{bmatrix}, C_2 = \begin{bmatrix} 0.3 & 0.4 \end{bmatrix}, H_2 = 0.3,$$

where a and b are varying elements, which can be used to compare the feasible areas for stability analysis. Suppose that a and b vary in the intervals $[-1, 1]$ and $[0, 15]$, respectively.

(A) Comparison of the feasible areas
Let us first consider the free switched system (5.1). Fig. 5.1 demonstrates the feasible areas by employing the common LF [12] (black diamond), nonmonotonic LF with $N = 2$ [27] (red star), and one-step ahead LF (blue circle).

It is clear from Fig. 5.1 that the proposed approach gives the largest feasible area.

(B) Comparison of the attenuation level γ
For $b = 2$ and a varying from 0 to 1.5, a comparison of the optimized attenuation level $\gamma_{\min}$ among the above-mentioned three approaches is given in Fig. 5.2. As expected, the one-step ahead LF approach yields the best optimized H_∞ performance.

(C) Time-difference of the energy function
For $a = 1$ and $b = 10.5$, Fig. 5.3 shows the evolution of the time-difference of the energy function, that is, the evolution of ΔV_k. Denoting $\Delta x_k = x_{k+1} - x_k$, we can see that ΔV_k (i.e., $\Delta x_k^T P \Delta x_k$) under the common LF method (black line) is less than 0, which implies that the energy of the system has a monotonic decreasing nature. The ΔV_k (i.e.,

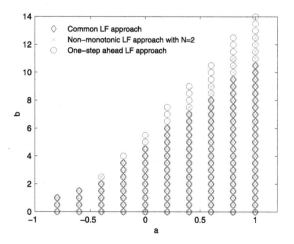

FIGURE 5.1   Feasible areas for the stability analysis.

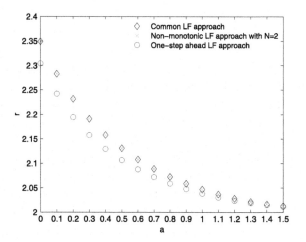

FIGURE 5.2 H_∞ performance.

$\Delta x_k^T(P_1 + P_2)\Delta x_k)$ of the nonmonotonic LF method with $N = 2$ (red line; light gray in print version) is larger than 0 at only three points, i.e., 0.0076 at the 7th step, 7.7865×10^{-4} at the 21th step, and 5.7899×10^{-5} at the 33rd step, which illustrates that such an LF is allowed to increase during the period of sampling time step and can reduce the conservatism in the stability criterion to some extent. The ΔV_k (i.e., $0.01 \times \Delta x_k^T P_{11} \Delta x_k$) of the one-step ahead LF approach (blue line; dark gray in print version) gives a more obvious nonmonotonic sign and therefore has greater potential in the conservatism reduction. The coefficient 0.01 is multiplied to make a clear comparison.

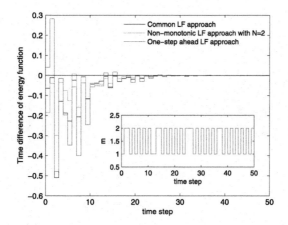

FIGURE 5.3 Evolution of the time difference.

TABLE 5.1 Comparison of the attenuation level.

	Stability	H_∞ performance
Common LF	3	4
NLF	6	10
One-step ahead LF	50	51

(D) Comparison of the decision variables

By observing Lemma 5.1 and Theorem 5.1 in this chapter the numbers of decision variables of different approaches are shown in Table 5.1. Obviously, the one-step ahead LF approach has more decision variables which is the cost to obtain a more flexible stability criterion.

Example 5.2. In this example, we borrow the system from [32], which has the following mathematical representation:

$$A_1 = \begin{bmatrix} 0.94 & 0.10 & 0.06 \\ -0.30 & 0.95 & -0.30 \\ -0.25 & -0.06 & 0.63 \end{bmatrix}, A_2 = \begin{bmatrix} 0.93 & 0.08 & 0.07 \\ -0.14 & 0.66 & -0.20 \\ -0.16 & -0.04 & 0.66 \end{bmatrix},$$

$$B_1 = \begin{bmatrix} -1.2 \\ 0.7 \\ -0.8 \end{bmatrix}, F_1 = \begin{bmatrix} -0.3 \\ 0.2 \\ 0.1 \end{bmatrix}, D_1 = 0,$$

$$C_1 = \begin{bmatrix} 0.7 & 0 & 0.3 \end{bmatrix}, H_1 = 0,$$

$$B_2 = \begin{bmatrix} -1.3 \\ -0.1 \\ 0 \end{bmatrix}, F_2 = \begin{bmatrix} -1.4 \\ -0.3 \\ 0.2 \end{bmatrix}, D_2 = 0,$$

$$C_2 = \begin{bmatrix} 0.2 & 0 & 0.4 \end{bmatrix}, H_2 = 0.$$

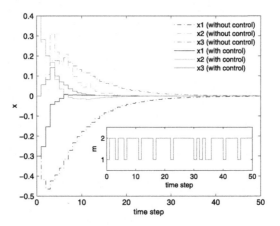

FIGURE 5.4 State trajectories of input-free system and closed-loop system.

The initial state and exogenous disturbance are taken as $[-0.3\ 0.4\ 0]^T$ and $w_k = 0.2e^{-0.5k}$. The state trajectories of the input-free system (5.1) and the closed-loop system (5.3) are drawn in Fig. 5.4. We can see that the transient response of the closed-loop system (solid lines) is much better than that of the input-free system (dash-dot lines). By solving the LMIs (5.25a) and (5.25b) presented in Theorem 5.3 we obtain the mode-independent controller gain $K = [0.6204\ 0.1502\ -0.1311]$ when the optimized attenuation level $\gamma_{\min}$ is reduced from 3.4135 to 1.2098.

5.6 CONCLUSION

In this chapter, we investigated the stability and H_∞ stabilization problems of discrete-time switched systems with arbitrarily switching signals. The concept of one-step ahead LF is introduced by constructing an augmented system with the knowledge of the future state one-step ahead of the current time, and therefore a less conservative stability criterion and better H_∞ attenuation capability can be satisfactorily achieved. Numerically testable stability and stabilization criteria are obtained via standard LMI formulation. A direct future work will be the development of the one-step ahead LF approach in the continuous-time domain or extending such an approach to the time-delayed switched systems.

References

[1] L.X. Zhang, T. Yang, P. Shi, Y.Z. Zhu, Analysis and Design of Markov Jump Systems with Complex Transition Probabilities, Springer, Switzerland, 2016.

[2] L.X. Zhang, Y.S. Leng, P. Colaneri, Stability and stabilization of discrete-time Semi-Markov jump linear systems via Semi-Markov Kernel approach, IEEE Transactions on Automatic Control 61 (2) (2016) 503–508.

[3] X.D. Zhao, L.X. Zhang, P. Shi, M. Liu, Stability of switched positive linear systems with average dwell time switching, Automatica 48 (6) (2012) 1132–1137.

[4] X.D. Zhao, L.X. Zhang, P. Shi, M. Liu, Stability and stabilization of switched linear systems with mode-dependent average dwell time, IEEE Transactions on Automatic Control 57 (7) (2012) 1809–1815.

[5] J.W. Wen, L. Peng, S.K. Nguang, Stochastic finite-time boundedness on switching dynamics Markovian jump linear systems with saturated and stochastic nonlinearities, Information Sciences 334 (2016) 65–82.

[6] J.W. Wen, S.K. Nguang, P. Shi, L. Peng, Finite-time stabilization of Markovian jump delay systems – a switching control approach, International Journal of Robust and Nonlinear Control 27 (2) (2017) 298–318.

[7] H.Q. Wang, P.X. Liu, S. Li, D. Wang, Adaptive neural output-feedback control for a class of non-lower triangular nonlinear systems with unmodeled dynamics, IEEE Transactions on Neural Networks and Learning Systems 29 (8) (2018) 3658–3668.

[8] H.Q. Wang, P.X. Liu, B. Niu, Robust fuzzy adaptive tracking control for non-affine stochastic nonlinear switching systems, IEEE Transactions on Cybernetics 48 (8) (2018) 2462–2471.

[9] Q.X. Zheng, H.B. Zhang, H_∞ filtering for a class of nonlinear switched systems with stable and unstable subsystems, Signal Processing 141 (2017) 240–248.

[10] C. Briat, Stability analysis and stabilization of stochastic linear impulsive, switched and sampled-data systems under dwell-time constraints, Automatica 74 (2) (2016) 279–287.

[11] X.C. Xie, J. Lam, P.S. Li, Finite-time H_∞ control of periodic piecewise linear systems, International Journal of Systems Science 48 (1) (2017) 2333–2344.

[12] R. Shorten, K.S. Narendra, O. Mason, A result on common quadratic Lyapunov functions, IEEE Transactions on Automatic Control 48 (1) (2003) 110–113.

[13] P. Mason, U. Boscain, Y. Chitour, Common polynomial Lyapunov functions for linear switched systems, SIAM Journal on Control and Optimization 45 (2006) 226–245.

[14] H. Lin, P.J. Antsaklis, Stability and stabilizability of switched linear systems: a survey of recent results, IEEE Transactions on Automatic Control 54 (2) (2009) 308–322.

[15] M.Z. Hou, F.Y. Fu, G.R. Duan, Global stabilization of switched stochastic nonlinear systems in strict-feedback form under arbitrary switchings, Automatica 49 (8) (2013) 2571–2575.

[16] B. Jiang, Q.K. Shen, P. Shi, Neural-networked adaptive tracking control for switched nonlinear pure-feedback systems under arbitrary switching, Automatica 61 (2005) 119–125.

[17] N. Sene, A. Chaillet, M. Balde, Relaxed conditions for the stability of switched nonlinear triangular systems under arbitrary switching, Systems & Control Letters 84 (2005) 52–56.

[18] S.F. Derakhshan, A. Fatehi, Non-monotonic Lyapunov functions for stability analysis and stabilization of discrete time Takagi–Sugeno fuzzy systems, International Journal of Innovative Computing, Information & Control (2014) 1567–1586.

[19] S.F. Derakhshan, A. Fatehi, M.G. Sharabiany, Nonmonotonic observer-based fuzzy controller designs for discrete time T-S fuzzy systems via LMI, IEEE Transactions on Cybernetics 44 (12) (2014) 2557–2567.

[20] S.F. Derakhshan, A. Fatehi, Non-monotonic robust H_2 fuzzy observer-based control for discrete time nonlinear systems with parametric uncertainties, International Journal of Systems Science 4 (12) (2015) 2134–2149.

[21] A. Kruszewski, T.M. Guerra, S. Labiod, Stabilization of Takagi–Sugeno discrete models: towards an unification of the results, in: IEEE International Fuzzy Systems Conference, 2007, pp. 1–6.

[22] A. Kruszewski, R. Wang, T.M. Guerra, Non-quadratic stabilization conditions for a class of uncertain nonlinear discrete time T-S fuzzy models: a new approach, IEEE Transactions on Automatic Control 53 (2) (2008) 606–611.

[23] T.M. Guerra, A. Kruszewski, S. Lauber, Discrete Tagaki–Sugeno models for control: where are we?, Annual Reviews in Control 33 (1) (2009) 37–47.

[24] T.M. Guerra, A. Kruszewski, M. Bernal, Control law proposition for the stabilization of discrete Takagi–Sugeno models, IEEE Transactions on Fuzzy Systems 17 (3) (2009) 724–731.

[25] Y.J. Chen, M. Tanaka, K. Inoue, H. Ohtake, K. Tanaka, T.M. Guerra, A. Kruszewski, H.O. Wang, A non-monotonically decreasing relaxation approach of Lyapunov functions to guaranteed cost control for discrete fuzzy systems, IET Control Theory & Applications 8 (16) (2014) 1716–1722.

[26] A. Nasiri, S.K. Nguang, A. Swain, D.J. Almakhles, Robust output feedback controller design of discrete-time Takagi–Sugeno fuzzy systems: a non-monotonic Lyapunov approach, IET Control Theory & Applications 10 (5) (2016) 545–553.

[27] A.A. Ahmadi, P.A. Parrilo, Non-monotonic Lyapunov functions for stability of discrete time nonlinear and switched systems, in: Proceedings of 47th IEEE Conference on Decision and Control, Cancun, Mexico, 2008, pp. 614–621.

[28] Y. Ebihara, D. Peaucelle, D. Arzelier, S-Variable Approach to LMI Based Robust Control, Springer, London, 2015.

[29] Y. Hosoe, D. Peaucelle, S-variable approach to robust stabilization state feedback synthesis for systems characterized by random polytopes, in: Proceedings of the 2016 European Control Conference, 2016, pp. 2023–2028.

[30] J.W. Wen, S.K. Nguang, P. Shi, N. Alireza, Robust H_∞ control of discrete-time non-homogenous Markovian jump systems via multi-step Lyapunov function approach, IEEE Transactions on Systems, Man, and Cybernetics: Systems 47 (7) (2017) 1439–1450.

[31] W.P. Dayawansa, C.F. Martin, A converse Lyapunov theorem for a class of dynamical systems which undergo switching, IEEE Transactions on Automatic Control 44 (4) (1999) 751–760.

[32] L. Zhang, N. Cui, M. Liu, Y. Zhao, Asynchronous filtering of discrete-time switched linear systems with average dwell time, IEEE Transactions on Circuits and Systems I 58 (5) (2011) 1109–1118.

Stability, l_2-Gain and Robust H_∞ Control of Switched Systems via Multistep Ahead Nonmonotonic Approach

6.1 INTRODUCTION

During the past decades, switched systems [1,2] have drawn an everincreasing attention to the control community due to their wide applications in many fields. To name a few, dc/dc switching converters [3], flight control [4], network control systems [5], etc. Generally, a typical switched system consists of a family of subsystems described by difference equations and a rule orchestrating the switching between them. It is worth mentioning that a new research trend in nowadays is studying the analysis and synthesis problem for nonlinear switched systems. For instance, the adaptive output-feedback control for a class of uncertain switched nonlinear stochastic systems in nonlower triangle form was investigated in detail [6]. Furthermore, a new adaptive approximation-based tracking controller design approach was developed in [7] by considering the output constraint.

As for switched systems, the switching law is crucial to the system dynamics; for example, even for a switched system including two unstable subsystems, we still can regulate it to be stable [8]. Therefore, it is desirable to develop an appropriate switching law to stabilize the system and achieve an improvement of performance. To this end, average dwell-time (ADT) switching [9,10], which means that the number of switches in a finite interval is bounded and the average time between two consecutive switches is not less than a specified value, has been demonstrated to be a flexible and effective switching logic for stability analysis and controller synthesis. Many advanced results for switched system with ADT switching have been reported in the literature. The stability problem for

91

switched positive linear systems was discussed in [11], and [12] proposed a systematic approach to the design of output-feedback switching linear parameter-varying (LPV) controllers with ADT switching. The fault detection and isolation problem was recently addressed in [13] for switched control systems in the presence of exogenous noise. In addition, combining the ADT switching method with Lyapunov function (LF) approach, [14] dealt with the model reduction problem for switched LPV systems by allowing the LF to increase at the switching instants. Taking a more practical phenomenon, i.e., the asynchronous switching, into account, [15] investigated the stability and H_∞ control problems of discrete-time switched systems, and the mode-dependent full-order filtering was further designed in [16], where the LFs are allowed a limited increase during the asynchronous mismatched period.

It is worth mentioning that ADT switching only allows the LF to increase at the switching instant and the asynchronous design places emphasis on the closed-loop synthesis, in which the maximum mismatched time need to be known a priori. Therefore, we believe a that new analysis and synthesis approach further considering the nonmonotonic nature within each subsystem still has great research potential from the conservatism reduction point of view.

Some pioneering works on the nonmonotonic Lyapunov function (NLF) have already been reported. The originality has come from the nonlinear and T-S fuzzy model. To reduce the conservatism of monotonicity requirement [17], local increase is allowed within each fuzzy rule, whereas the time-difference of the LF ultimately converges to zero. Two-sample variation, i.e., $V(x_{k+2}) < V(x_k)$ [18], and N-sample variation, i.e., $V(x_{k+N}) < V(x_k)$ [19], were introduced to demonstrate the advantages of such a nonmonotonic approach. Subsequently, a more general LF approach, i.e., $\sum_{i=1}^{N} (V_i(x_{k+i}) - V_i(x_k)) < 0$, was proposed in [20]. Meanwhile, great efforts have been devoted to extending the NLF approach so that less conservative results could be obtained for various dynamical systems. For instance, based on two-sample variation, an observer-based fuzzy controller [21] and robust H_2 observer-based controller [22] were further introduced by combining with a parallel-distributed compensation scheme. By extending the N-step approach, new stabilizable conditions with further less conservatism were presented in [23,24], and a new guaranteed cost control design was also developed in [25]. Based on the fundamental work of [20], the multistep LF approach was extended to a class of inhomogenous Markovian jump systems, which allows an LF to increase during the period of several sampling time steps ahead of the current time within the jump mode [26]. A similar method was adopted to reduce conservatism in H_∞ robust state feedback and output feedback controller design of T-S fuzzy systems [27,28]. Despite the popularity of the investigation on the NLF approach, the ADT switching mechanism has been largely overlooked due probably to the mathematical complexity, especially when the N-step ahead scenario is considered, which motivates our present research.

Based on the above discussions, the aim of this chapter is to solve the robust H_∞ control problem via an N-step ahead LF approach for discrete-time switched systems under the ADT design framework. The predictive horizon and switching interval $[k_l, k_{l+1}]$ are considered to construct an exponential damping law of the decreasing points of LF. We establish testable conditions under which the closed-loop system is asymptotically stable and achieves an optimized H_∞ attenuation level. The main contributions and novelty of this chapter are summarized as follows:

1) The N-step ahead approach, i.e., $V(x_{k+N}) - V(x_k) < 0$, is introduced for switched systems with ADT switching. Therefore, an LF is allowed to increase both at the switching instant and within each subsystem, and the direct benefit is enlarging the stability region.

2) The l_2-gain is analyzed by moving the horizon from $k + N - 1$ to k step by step. Auxiliary matrices are introduced to avoid constructing large matrix inequalities instead of testing a set of linear matrix inequalities (LMIs). The simulation results demonstrate that our approach achieves a better disturbance attenuation performance.

3) The relationship between the N-step time difference of LF and ADT constraints is thoroughly discussed, i.e., the switching must be gentle when we obtain advantages addressed in 1) and 2).

The layout of the remainder is as follows. In Section 6.2, we briefly describe the system and problem formulation. By employing the N-step ahead LF approach we present a stability and l_2-gain analysis in Section 6.3. In Section 6.4, we formulate the problem of a robust H_∞ control. A practical example is given in Section 6.5 to demonstrate the potential of the proposed techniques, and conclusions are drawn in Section 6.6.

Notations: Throughout this chapter, the superscript "T" stands for matrix transposition, $\mathbb{R}^n$ is the n-dimensional Euclidean space, $l_2[0, \infty]$ is the space of square-summable infinite sequences, and for $\omega = \{\omega(k)\} \in l_2[0, \infty]$, its norm is given by $\|\omega\|_2 = \sqrt{\sum_{k=0}^\infty |\omega(k)|^2}$. Furthermore, a function $\beta : [0, \infty) \times [0, \infty) \to [0, \infty)$ is said to be of class $\mathcal{KL}$ if $\beta(\cdot, t)$ is of class $\mathcal{K}$ for each fixed $t \geqslant 0$ and $\beta(s, t)$ decreases to 0 as $t \to \infty$ for each fixed $s \geqslant 0$. The symbol $*$ is used in block matrices as an ellipsis for the symmetric terms. We let $diag\{\cdots\}$ stand for a block-diagonal matrix and $P > 0$ ($P \geqslant 0$) denote a positive definite (positive semidefinite) matrix; $\aleph_{r_0, r} \triangleq \{r_0, \ldots, r\}$, where $r_0, \ldots, r \in \mathbb{Z}^+$, and $\mathbb{Z}^+$ is the set of nonnegative integers.

6.2 PRELIMINARIES AND PROBLEM FORMULATION

Consider the class of discrete-time uncertain switched systems given by

$$\begin{cases} x_{k+1} = (A_{\sigma(k)} + \Delta A^\eta_{\sigma(k)})x_k + (B_{\sigma(k)} + \Delta B^\eta_{\sigma(k)})u_k + B_{\sigma(k)\omega}\omega_k, \\ z_k = C_{\sigma(k)}x_k + D_{\sigma(k)}u_k, \end{cases} \tag{6.1}$$

where $x_k \in \mathbb{R}^n$, u_k, $z_k \in \mathbb{R}^p$, and $\omega_k \in \mathbb{R}^q$ are the discrete state vector of the system, control input vector, the controlled output vector, and disturbance input belonging to $l_2[0, \infty]$, respectively; $\sigma(k)$ is called a switching signal, which takes values in the finite set $\aleph_{1,m}$. For each possible value of $\sigma(k) = i$, the constant matrices A_i, B_i, C_i, $B_{i\omega}$, and D_i are of appropriate dimensions, and the uncertainties are assumed in the following form:

$$[\Delta A^\eta_i \ \ \Delta B^\eta_i] = S_i F_\eta [H_i \ \ N_i], \tag{6.2}$$

where S_i, H_i, and N_i are known constant matrices of appropriate dimensions, and F_η, $\eta \in [k, k + N]$, are unknown time-varying matrix functions satisfying $F^T_\eta F_\eta < I$.

In this chapter, we consider the control input in system (6.1) with $u_k = K_i x_k$, where K_i is the controller gain to be determined. For convenience, we set $\tilde{A}^\eta_{\sigma(k)} \triangleq A_{\sigma(k)} + \Delta A^\eta_{\sigma(k)}$, $\tilde{B}^\eta_{\sigma(k)} \triangleq B_{\sigma(k)} + \Delta B^\eta_{\sigma(k)}$, $\bar{A}_{\sigma(k)} \triangleq \tilde{A}^\eta_{\sigma(k)} + \tilde{B}^\eta_{\sigma(k)} K_i$, and $\bar{C}_{\sigma(k)} \triangleq C_{\sigma(k)} + D_{\sigma(k)} K_i$, and thus (6.1) can be rewritten as follows:

$$\begin{cases} x_{k+1} = \bar{A}_{\sigma(k)} x_k + B_{\sigma(k)\omega}\omega_k, \\ z_k = \bar{C}_{\sigma(k)} x_k. \end{cases} \tag{6.3}$$

Now we recall the definitions of asymptotical stability and ADT switching for system (6.1) along with a useful lemma for further development.

Definition 6.1. [12] System (6.1) is said to be globally uniformly asymptotically stable (GUAS) under a switching signal $\sigma(k)$ if there exists a function β of class $\mathcal{KL}$ such that the solution x_k of the system satisfies $\|x(k)\| \le \beta(\|x(k_0)\|, k)$ for $k > k_0$.

Definition 6.2. [14] For switching signal $\sigma(k)$ and any $K > k > k_0$, let $N_\sigma(K, k)$ stand for the switching numbers of σ over the interval $[k, K)$. If $N_\sigma(K, k) \le N_0 + (K - k)/\tau_a$ for any given $N_0 > 0$ and $\tau_a > 0$, then τ_a is referred to as ADT, and N_0 is the chatter bound.

Lemma 6.1. *[19] For constant matrices Y, H, and E of appropriate dimensions and any time-varying matrix F_k satisfying $F^T_k F_k < I$, we have the inequality $Y + HF_k E + (HF_k E)^T < 0$ if and only if there exists a scalar $\varepsilon > 0$ such that $Y + \varepsilon HH^T + \varepsilon^{-1} E^T E < 0$.*

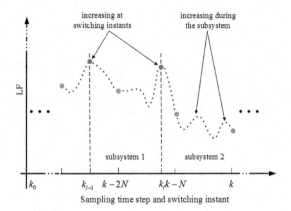

FIGURE 6.1 LF evolution under ADT switching.

The LF evolution under ADT switching via N-step ahead approach is depicted in Fig. 6.1. We assume that the switching behavior is much slower than sampling. The values of LF at the switching instants $k_l, k_{l+1} \cdots k_{l+m-1}$ are marked with green solid dots (dark gray in print version). Red solid dots (light gray in print version) are the values of decreasing point within the subsystem in terms of the N-step ahead approach. By observing Fig. 6.1 it is clear that the ADT switching allows the LF to increase at the switching instants whereas the N-step ahead approach allows LF to increase within the subsystems.

6.3 STABILITY AND l_2-GAIN ANALYSIS

In this section, we first formulate, via a class of N-step ahead LFs, the relaxed stability criterion and the minimal ADT constraint for system (6.1) to be GUAS without considering uncertainty and disturbance. Then, based on the stability results, we further analyze the l_2-gain.

Theorem 6.1. *Given constants $0 < \alpha < 1$ and $\mu > 1$, the switched system (6.1) without uncertainties and disturbance is GUAS if there exists a set of symmetric matrices $Q_{i,f}$ ($f \in \aleph_{1,N-1}$) and $P_i > 0$ such that*

for $N = 1$,

$$A_i^T P_i A_i + (\alpha - 1)P_i \le 0, \tag{6.4}$$

for $N = 2$,

$$A_i^T P_i A_i - Q_{i,1} \le 0, \tag{6.5}$$

$$A_i^T Q_{i,1} A_i + (\alpha - 1)P_i \le 0, \tag{6.6}$$

for $N \geq 3$,

$$A_i^T P_i A_i - Q_{i,N-1} \leq 0, \tag{6.7}$$

$$A_i^T Q_{1,N-f+1} A_i - Q_{i,N-f} \leq 0, f \in \aleph_{2,N-1}, \tag{6.8}$$

$$A_i^T Q_{i,1} A_i + (\alpha - 1) P_i \leq 0, \tag{6.9}$$

and

$$P_i \leq \mu P_j \tag{6.10}$$

for all $i, j \in \Omega$, and the ADT of the switching signal $\sigma(k)$ satisfies

$$\tau_a > \tau_a^* = -N \ln \mu / \ln(1 - \alpha). \tag{6.11}$$

Proof. We consider a class of quadratic LFs given by

$$V_i(x_k) = x_k^T P_i x_k. \tag{6.12}$$

Take the N-step time difference of such an LF as follows:

$$\Delta V_{i,N}(x_k) = V_i(x_{k+N}) - V_i(x_k). \tag{6.13}$$

Let

$$V_i'(x_{k+f}) = x_{k+f}^T Q_{i,f} x_{k+f}, f \in \aleph_{1,N-1}. \tag{6.14}$$

Step 1: When $N \geq 3$, adding and subtracting $V_i'(x_{k+N-1})$ to and from (6.13), we have

$$\Delta V_{i,N}(x_k) = V_i(x_{k+N}) + V_i'(x_{k+N-1}) - V_i'(x_{k+N-1}) - V_i(x_k). \tag{6.15}$$

Using $x_{k+N} = A_i x_{k+N-1}$, we obtain

$$\Delta V_{i,N}(x_k) = x_{k+N-1}^T Q_{i,N-1} x_{k+N-1} - x_k^T P_i x_k$$
$$+ x_{k+N-1}^T (A_i^T P_i A_i - Q_{i,N-1}) x_{k+N-1}. \tag{6.16}$$

If condition (6.7) holds, then

$$\Delta V_{i,N}(x_k) \leq x_{k+N-1}^T Q_{i,N-1} x_{k+N-1} - x_k^T P_i x_k. \tag{6.17}$$

Step 2: Adding and subtracting $V_i'(x_{k+N-2})$ to and from (6.17) and bearing in mind that $x_{k+N-1} = A_i x_{k+N-2}$, we have

$$\Delta V_{i,N}(x_k) \leq x_{k+N-2}^T (A_i^T Q_{i,N-1} A_i - Q_{i,N-2}) x_{k+N-2}$$
$$+ x_{k+N-2}^T Q_{i,N-2} x_{k+N-2} - x_k^T P_i x_k. \tag{6.18}$$

If condition (6.8) holds for $f = 2$, then

$$x_{k+N-2}^T (A_i^T Q_{i,N-1} A_i - Q_{i,N-2}) x_{k+N-2} \leq 0$$

leads to

$$\Delta V_{i,N}(x_k) \leq x_{k+N-2}^T Q_{i,N-2} x_{k+N-2} - x_k^T P_i x_k. \tag{6.19}$$

Repeating the procedure $N - 3$ times, we have

$$\Delta V_{i,N}(x_k) \leq x_{k+1}^T Q_{i,1} x_{k+1} - x_k^T P_i x_k. \tag{6.20}$$

Step 3: Adding and subtracting $\alpha V_i(x_k)$ to and from (6.20) and employing $x_{k+1} = A_i x_k$ yield

$$\Delta V_{i,N}(x_k) \leq x_k^T \left(A_i^T Q_{i,1} A_i + (\alpha - 1) P_i \right) x_k - \alpha x_k^T P_i x_k. \tag{6.21}$$

If condition (6.9) holds, then this inequality results in $\Delta V_{i,N}(x_k) \leq -\alpha V_i(x_k)$. When $N = 2$, replacing N of Step 1 by 2 and then passing directly to Step 3, we easily infer $\Delta V_{i,2}(x_k) \leq -\alpha V_i(x_k)$ from conditions (6.5) and (6.6).

Similarly, when $N = 1$, performing Step 3 directly when condition (6.4) holds, we can obtain that $\Delta V_{i,1}(x_k) \leq -\alpha V_i(x_k)$. Therefore, for all $N \in \mathbb{Z}^+$, the condition

$$V_i(x_{k+N}) - V_i(x_k) \leq -\alpha V_i(x_k) \tag{6.22}$$

holds.

With the assumption that the switching behavior is much slower than sampling time step, we have

$$V_{\sigma(k)}(x_k) \leq (1 - \alpha)^{\frac{k-k_l}{N}} V_{\sigma(k)}(x_{k_l}). \tag{6.23}$$

By condition (6.10) we have

$$V_i(x_{k_l}) \leq \mu V_j(x_{k_l}). \tag{6.24}$$

By substituting (6.20) into (6.19) we obtain

$$V_{\sigma(k)}(x_k) \leq \mu(1 - \alpha)^{\frac{k-k_l}{N}} V_{\sigma(k_l)}(x_{k_l}),$$
$$V_{\sigma(k)}(x_k) \leq \mu(1 - \alpha)^{\frac{k-(k_l-1)}{N}} V_{\sigma(k_l)}(x_{k_l-1}),$$
$$\vdots$$
$$V_{\sigma(k)}(x_k) \leq \mu^2(1 - \alpha)^{\frac{k-k_{l+1}}{N}} V_{\sigma(k_{l+1})}(x_{k_{l+1}}),$$
$$\vdots$$

$$V_{\sigma(k)}(x_k) \le \mu^{N_0 + \frac{k-k_0}{\tau_a}} (1-\alpha)^{\frac{k-k_0}{N}} V_{\sigma(k_0)}(x_{k_0}),$$
$$V_{\sigma(k)}(x_k) \le \mu^{N_0} [\mu^{\frac{1}{\tau_a}}(1-\alpha)^{\frac{1}{N}}]^{k-k_0} V_{\sigma(k_0)}(x_{k_0}). \qquad (6.25)$$

If the ADT satisfies (6.11), then we have

$$(1-\alpha)^{1/N} \mu^{1/\tau_a} = e^{\ln((1-\alpha)^{1/N}) + \ln(\mu^{1/\tau_a})} = e^0 = 1. \qquad (6.26)$$

Consequently, we can derive $V_{\sigma(k)}(x_k) \le \mu^{N_0} V_{\sigma(k_0)}(x_{k_0})$, and thus the GUAS of system (6.1) without uncertainty and disturbance is established according to Definition 6.1. The proof is completed. $\qquad \square$

Remark 6.1. The matrices $Q_{i,f}$ are introduced as auxiliary matrices such that large matrix inequalities can be avoided by constructing a set of LMIs, especially when dealing with l_2-gain analysis. Note that there are no positive constraints on the $Q_{i,f}$, and therefore they can be viewed as slack variables to further reduce the conservatism of the stability criterion. Moreover, by observing the ADT constraint (6.11) we can conclude that increasing the number N can give a larger stability region along with a gentler requirement on the switching rate.

Remark 6.2. As can be seen in (6.23), due to the nonmonotonic nature, although LF does not decrease at every sampling time, it still has an average decrease on every N sampling step in an exponential sense. Therefore, a possible increase is allowed at several sampling steps ahead of the current time, and a less conservative criterion can be expected.

Now, by further invoking the nonmonotonic LF addressed in the previous theorem the l_2-gain analysis for system (6.1) without uncertainty is given in the following theorem.

Theorem 6.2. *Given some scalars $0 < \alpha < 1$ and $\mu > 1$, suppose that there exist a scalar $\gamma > 0$, a set of symmetric matrices $Q_{i,f}$ ($f \in \aleph_{1,N-1}$), and a symmetric positive matrix P_i such that*

for $N = 1$,

$$\begin{bmatrix} \Omega_1^1 & A_i^T P_i B_{i\omega} \\ * & -\gamma^2 I + B_{i\omega}^T P_i B_{i\omega} \end{bmatrix} \le 0, \qquad (6.27)$$

for $N = 2$,

$$\begin{bmatrix} \Omega_2^1 & A_i^T P_i B_{i\omega} \\ * & -\gamma^2 I + B_i^T P_i B_{i\omega} \end{bmatrix} \le 0, \qquad (6.28)$$

$$\begin{bmatrix} \Omega_N^N & A_i^T Q_{i,1} B_{i\omega} \\ * & -\gamma^2 I + B_i^T Q_{i,1} B_{i\omega} \end{bmatrix} \le 0, \qquad (6.29)$$

for $N \geq 3$,

$$\begin{bmatrix} \Omega_N^1 & A_i^T P_i B_{i\omega} \\ * & -\gamma^2 I + B_i^T P_i B_{i\omega} \end{bmatrix} \leq 0, \tag{6.30}$$

$$\begin{bmatrix} \Omega_N^f & A_i^T Q_{i,N-f+1} B_{i\omega} \\ * & -\gamma^2 I + B_i^T Q_{i,N-f+1} B_{i\omega} \end{bmatrix} \leq 0, \tag{6.31}$$

$$\begin{bmatrix} \Omega_N^N & A_i^T Q_{i,1} B_{i\omega} \\ * & -\gamma^2 I + B_i^T Q_{i,1} B_{i\omega} \end{bmatrix} \leq 0, \tag{6.32}$$

where

$$\Omega_1^1 = (\alpha - 1) P_i + A_i^T P_i A_i + C_i^T C_i,$$
$$\Omega_2^1 = -Q_{i,1} + A_i^T P_i A_i + C_i^T C_i,$$
$$\Omega_N^1 = -Q_{i,N-1} + A_i^T P_i A_i + C_i^T C_i,$$
$$\Omega_N^f = -Q_{i,N-f} + A_i^T Q_{i,N-f+1} A_i + C_i^T C_i, \ f \in \aleph_{2,N-1},$$
$$\Omega_N^N = (\alpha - 1) P_i + A_i^T Q_{i,1} A_i + C_i^T C_i,$$

$$P_i \leq \mu P_j \tag{6.33}$$

for all $i, j \in \Omega$, and the ADT of the switching signal $\sigma(k)$ satisfies

$$\tau_a > \tau_a^* = -N \ln \mu / \ln(1 - \alpha). \tag{6.34}$$

Then system (6.1) without uncertainty is GUAS with the l_2-gain γ.

Proof. By (6.27)–(6.34) it is clear that Theorem 6.2 implies the GUAS. Define the l_2-gain index as

$$J_M \overset{\Delta}{=} \| z \|_2^2 - \gamma^2 \| \omega \|_2^2 = \sum_{k=0}^{M-1} \{ z_k^T z_k - \gamma^2 \omega_k^T \omega_k \}. \tag{6.35}$$

Next, we pay our attention to the l_2-gain analysis.

Step 1: When $N \geq 3$, adding and subtracting $V_i'(x_{k+N-1}) + z_{k+N-1}^T z_{k+N-1} - \gamma^2 \omega_{k+N-1}^T \omega_{k+N-1}$ to and from (6.17), we get

$$\begin{aligned} \Delta V_{i,N}(x_k) = {}& V_i(x_{k+N}) - V_i(x_k) + V_i'(x_{k+N-1}) \\ & - V_i'(x_{k+N-1}) + z_{k+N-1}^T z_{k+N-1} \\ & - \gamma^2 \omega_{k+N-1}^T \omega_{k+N-1} - (z_{k+N-1}^T z_{k+N-1} \\ & - \gamma^2 \omega_{k+N-1}^T \omega_{k+N-1}). \end{aligned} \tag{6.36}$$

Using $x_{k+N} = A_i x_{k+N-1} + B_{i\omega}\omega_{k+N-1}$ and $z_{k+N-1} = C_i x_{k+N-1}$, we have

$$
\begin{aligned}
\Delta V_{i,N}(x_k) = {} & \omega_{k+N-1}^T (B_i^T P_i B_{i\omega} - \gamma^2 I)\omega_{k+N-1} \\
& + x_{k+N-1}^T (A_i^T P_i A_i - Q_{i,N-1} + C_i^T C_i) \\
& \times x_{k+N-1} + 2x_{k+N-1}^T (A_i^T P_i B_{i\omega})\omega_{k+N-1} \\
& - (z_{k+N-1}^T z_{k+N-1} - \gamma^2 \omega_{k+N-1}^T \omega_{k+N-1}) \\
& + x_{k+N-1}^T Q_{i,N-1} x_{k+N-1} - x_k^T P_i x_k.
\end{aligned} \tag{6.37}
$$

From condition (6.30) it follows that

$$
\begin{aligned}
\Delta V_{i,N}(x_k) \le {} & -(z_{k+N-1}^T z_{k+N-1} - \gamma^2 \omega_{k+N-1}^T \omega_{k+N-1}) \\
& + x_{k+N-1}^T Q_{i,N-1} x_{k+N-1} - x_k^T P_i x_k.
\end{aligned} \tag{6.38}
$$

Step 2: Adding and subtracting $z_{k+N-2}^T z_{k+N-2} - \gamma^2 \omega_{k+N-2}^T \omega_{k+N-2} + V_i'(x_{k+N-2})$ to and from (6.38) and bearing in mind that $x_{k+N-1} = A_i x_{k+N-2} + B_{i\omega}\omega_{k+N-2}$ and $z_{k+N-2} = C_i x_{k+N-2}$, we have

$$
\begin{aligned}
\Delta V_{i,N}(x_k) = {} & x_{k+N-2}^T Q_{i,N-2} x_{k+N-2} \\
& + \omega_{k+N-2}^T (B_i^T Q_{i,N-2} B_{i\omega} - \gamma^2 I)\omega_{k+N-2} \\
& + x_{k+N-2}^T (A_i^T Q_{i,N-1} A_i - Q_{i,N-2} \\
& + C_i^T C_i)x_{k+N-2} - \sum_{f=N-2}^{N-1} (z_{k+f}^T z_{k+f} \\
& - \gamma^2 \omega_{k+f}^T \omega_{k+f}) - x_k^T P_i x_k.
\end{aligned} \tag{6.39}
$$

If condition (6.31) holds for $f = 2$, this implies that

$$
\begin{aligned}
\Delta V_{i,N}(x_k) \le {} & x_{k+N-2}^T Q_{i,N-2} x_{k+N-2} - x_k^T P_i x_k \\
& - \sum_{f=N-2}^{N-1} (z_{k+f}^T z_{k+f} - \gamma^2 \omega_{k+f}^T \omega_{k+f}).
\end{aligned} \tag{6.40}
$$

Repeating the procedure $N - 3$ times, we obtain that

$$
\begin{aligned}
\Delta V_{i,N}(x_k) \le {} & x_{k+1}^T Q_{i,1} x_{k+1} - x_k^T P_i x_k \\
& - \sum_{f=1}^{N-1} (z_{k+f}^T z_{k+f} - \gamma^2 \omega_{k+f}^T \omega_{k+f}).
\end{aligned} \tag{6.41}
$$

Step 3: Adding and subtracting $\alpha V_i(x_k) + z_k^T z_k - \gamma^2 \omega_k^T \omega_k$ to and from (6.41) and employing $x_{k+1} = A_i x_k$ yield

$$\Delta V_{i,N}(x_k) \le x_k^T (A_i^T Q_{i,1} A_i + (\alpha - 1) P_i$$
$$+ C_i^T C_i) x_k + 2 x_k^T (A_i^T Q_{i,1} B_{i\omega}) \omega_k$$
$$+ \omega_k^T (B_i^T Q_{i,1} B_{i\omega} - \gamma^2 I) \omega_k - \alpha V_i(x_k) \tag{6.42}$$
$$- \sum_{f=0}^{N-1} (z_{k+f}^T z_{k+f} - \gamma^2 \omega_{k+f}^T \omega_{k+f}).$$

From (6.32) we have that

$$\sum_{f=0}^{N-1} (z_{k+f}^T z_{k+f} - \gamma^2 \omega_{k+f}^T \omega_{k+f}) + \Delta V_{i,N}(x_k) \le -\alpha V_i(x_k). \tag{6.43}$$

For $N = 2$, performing Step 1 and then directly Step 3, we can infer (6.43) from conditions (6.28) and (6.29).

For $N = 1$, implementing Step 3 directly and supposing that condition (6.27) holds, we can obtain condition (6.43). Then, taking the summation from $k = -N + 1$ to $k = M$ on both sides of (6.43), we have

$$\sum_{k=-N+1}^{M} \left\{ \sum_{f=0}^{N-1} (z_{k+f}^T z_{k+f} - \gamma^2 \omega_{k+f}^T \omega_{k+f}) \right\} + \sum_{k=-N+1}^{M} \{\Delta V_{i,N}(x_k)\}$$
$$\le \sum_{k=-N+1}^{M} \{-\alpha V_i(x_k)\}. \tag{6.44}$$

Under the zero initial conditions $x_0 = x_{-1} = x_{-2} \cdots = x_{-N+1} = 0$, it follows that

$$\sum_{k=-N+1}^{M} \{\Delta V_{i,N}(x_k)\} = V_{i,M+1} + V_{i,M+2} + \cdots + V_{i,M+N}. \tag{6.45}$$

As $M \to \infty$, we have that

$$N V_{i,\infty} + N J_\infty \le \sum_{k=1}^{\infty} \{-\alpha V_i(x_k)\}. \tag{6.46}$$

Since $N V_{i,\infty} \ge 0$ and $\sum_{k=1}^{\infty} \{-\alpha V_i(x_k)\} < 0$, we can conclude that $J_M < 0$, i.e., $\| z \|_2^2 < \gamma^2 \| \omega \|_2^2$, which means that the prescribed disturbance attenuation level is obtained. The proof is completed. $\qquad\square$

6.4 ROBUST H_∞ CONTROL

In this section, we provide sufficient conditions for the existence of a robust state feedback H_∞ controller.

Theorem 6.3. *Let $0 < \alpha < 1$ and $\mu > 1$. Suppose that there exist a set of symmetric matrices $Q_{i,f}$ ($f \in \aleph_{1.N-1}$), $P_i > 0$, matrices G and M_i of appropriate dimension, constant values $\varepsilon_i > 0$ and $\gamma > 0$ such that*

for $N = 1$,

$$
\begin{bmatrix}
(\alpha - 1)P_i & * & * & * & * \\
0 & -\gamma^2 I & 0 & * & * \\
A_i G + B_i M_i & B_{i\omega} & \Theta_1 & * & * \\
C_i G + D_i M_i & 0 & 0 & -I & * \\
H_i G + N_i M_i & 0 & 0 & 0 & -\varepsilon_i I
\end{bmatrix} \le 0,
\tag{6.47}
$$

for $N = 2$,

$$
\begin{bmatrix}
-Q_{i,1} & * & * & * & * \\
0 & -\gamma^2 I & 0 & * & * \\
A_i G + B_i M_i & B_{i\omega} & \Theta_1 & * & * \\
C_i G + D_i M_i & 0 & 0 & -I & * \\
H_i G + N_i M_i & 0 & 0 & 0 & -\varepsilon_i I
\end{bmatrix} \le 0,
\tag{6.48}
$$

$$
\begin{bmatrix}
(\alpha - 1)P_i & * & * & * & * \\
0 & -\gamma^2 I & 0 & * & * \\
A_i G + B_i M_i & B_{i\omega} & \Theta_N & * & * \\
C_i G + D_i M_i & 0 & 0 & -I & * \\
H_i G + N_i M_i & 0 & 0 & 0 & -\varepsilon_i I
\end{bmatrix} \le 0,
\tag{6.49}
$$

for $N \ge 3$,

$$
\begin{bmatrix}
-Q_{i,N-1} & * & * & * & * \\
0 & -\gamma^2 I & 0 & * & * \\
A_i G + B_i M_i & B_{i\omega} & \Theta_1 & * & * \\
C_i G + D_i M_i & 0 & 0 & -I & * \\
H_i G + N_i M_i & 0 & 0 & 0 & -\varepsilon_i I
\end{bmatrix} \le 0,
\tag{6.50}
$$

$$
\begin{bmatrix}
-Q_{i,N-f} & * & * & * & * \\
0 & -\gamma^2 I & 0 & * & * \\
A_i G + B_i M_i & B_{i\omega} & \Theta_f & * & * \\
C_i G + D_i M_i & 0 & 0 & -I & * \\
H_i G + N_i M_i & 0 & 0 & 0 & -\varepsilon_i I
\end{bmatrix} \leq 0,
\tag{6.51}
$$

$$
\begin{bmatrix}
(\alpha - 1) P_i & * & * & * & * \\
0 & -\gamma^2 I & 0 & * & * \\
A_i G + B_i M_i & B_{i\omega} & \Theta_N & * & * \\
C_i G + D_i M_i & 0 & 0 & -I & * \\
H_i G + N_i M_i & 0 & 0 & 0 & -\varepsilon_i I
\end{bmatrix} \leq 0,
\tag{6.52}
$$

where

$$
\Theta_1 = -G^T - G + P_i + \varepsilon_i S_i S_i^T,
$$
$$
\Theta_f = -G^T - G + Q_{i,N-f+1} + \varepsilon_i S_i S_i^T, \, f \in \aleph_{2.N-1},
$$
$$
\Theta_N = -G^T - G + Q_{i,1} + \varepsilon_i S_i S_i^T,
$$

$$
P_i \leq \mu P_j
\tag{6.53}
$$

for all $\forall i, j \in \Omega$, and the ADT of the switching signal $\sigma(k)$ satisfies

$$
\tau_a > \tau_a^* = -N \ln \mu / \ln(1 - \alpha).
\tag{6.54}
$$

Then, under the controller $u_k = M_i G^{-1} x_k$, system (6.1) is robust GUAS with the l_2-gain γ.

Proof. Consider the LF with symmetric matrices $P_i > 0$ and a matrix G given by

$$
V_i(x_k) = x_k^T G^{-T} P_i G^{-1} x_k.
\tag{6.55}
$$

Similarly to the proof of Theorem 6.2 for $N \geq 3$ and replacing $V_i(x_k)$ with the form of (6.55), we obtain sufficient conditions for robust stability:

$$
\begin{cases}
\Xi_1 = \begin{bmatrix} \Xi_1^{11} & \bar{A}_i^T G^{-T} P_i G^{-1} B_{i\omega} \\ * & \Xi_1^{22} \end{bmatrix} \leq 0, \\[2mm]
\Xi_f = \begin{bmatrix} \Xi_f^{11} & \bar{A}_i^T G^{-T} Q_{i,N-f+1} G^{-1} B_{i\omega} \\ * & \Xi_f^{22} \end{bmatrix} \leq 0, \\[2mm]
\Xi_N = \begin{bmatrix} \Xi_N^{11} & \bar{A}_i^T G^{-T} Q_{i,1} G^{-1} B_{i\omega} \\ * & \Xi_N^{22} \end{bmatrix} \leq 0,
\end{cases}
\tag{6.56}
$$

where

$$\Xi_1^{11} = -G^{-T}Q_{i,N-1}G^{-1} + \bar{A}_i^T G^{-T} P_i G^{-1}\bar{A}_i + \bar{C}_i^T\bar{C}_i,$$

$$\Xi_1^{22} = -\gamma^2 I + B_{i\omega}^T G^{-T} P_i G^{-1}{}_i B_{i\omega},$$

$$\Xi_f^{11} = -G^{-T}Q_{i,N-f}G^{-1} + \bar{C}_i^T\bar{C}_i + \bar{A}_i^T G^{-T}Q_{i,N-f+1}G^{-1}\bar{A}_i,$$

$$\Xi_f^{22} = -\gamma^2 I + B_{i\omega}^T G^{-T}Q_{i,N-f+1}G^{-1}{}_i B_{i\omega},$$

$$\Xi_N^{11} = (\alpha-1)G^{-T}P_i G^{-1} + \bar{A}_i^T G^{-T}Q_{i,1}G^{-1}\bar{A}_i + \bar{C}_i^T\bar{C}_i,$$

$$\Xi_N^{22} = -\gamma^2 I + B_{i\omega}^T G^{-T}Q_{i,1}G^{-1}{}_i B_{i\omega}.$$

Performing the congruence transformation to Ξ_1 with $diag\{G^T, I, I\}$ and $diag\{G, I, I\}$, noting that $-G^T - G + P_i \leq -GP_i^{-1}G^T$, setting $M_i \triangleq K_i G$, and employing Schur's complement, we obtain

$$
\begin{bmatrix}
-Q_{i,N-1} & * & * & * \\
0 & -\gamma^2 I & * & * \\
\tilde{A}_i^{k+N-1}G + \tilde{B}_i^{k+N-1}M_i & B_{i\omega} & -G^T - G + P_i & * \\
C_i G + D_i M_i & 0 & 0 & I
\end{bmatrix} \leq 0. \qquad (6.57)
$$

Subsequently, (6.57) can be decomposed as

$$
\begin{bmatrix}
-Q_{i,N-1} & * & * & * \\
0 & -\gamma^2 I & * & * \\
A_i G + B_i M_i & B_{i\omega} & -G^T - G + P_i & * \\
C_i G + D_i M_i & 0 & 0 & -I
\end{bmatrix}
$$

$$
+ \left\{
\begin{bmatrix} 0 \\ 0 \\ S_i \\ 0 \end{bmatrix}
F_{k+N-1}\begin{bmatrix} H_i G + N_i M_i & 0 & 0 & 0 \end{bmatrix}
\right\}^T \qquad (6.58)
$$

$$
+ \begin{bmatrix} 0 \\ 0 \\ S_i \\ 0 \end{bmatrix}
F_{k+N-1}\begin{bmatrix} H_i G + N_i M_i & 0 & 0 & 0 \end{bmatrix} \leq 0.
$$

By applying Lemma 6.1 from (6.58) it follows that

$$
\begin{bmatrix}
\Upsilon_1 & * & * & * \\
0 & -\gamma^2 I & * & * \\
A_i G + B_i M_i & B_{i\omega} & \Upsilon_2 & * \\
C_i G + D_i M_i & 0 & 0 & -I
\end{bmatrix} \leq 0, \qquad (6.59)
$$

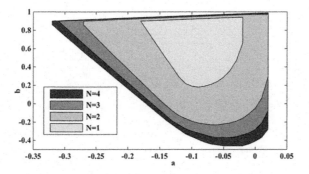

FIGURE 6.2 Comparison on the stability regions.

where $\Upsilon_1 = -Q_{i,N-1} + \varepsilon_i^{-1}(H_i G + N_i M_i)^T (H_i G + N_i M_i)$ and $\Upsilon_2 = -G^T - G + P_i + \varepsilon_i S_i S_i^T$, which directly implies that condition (6.50) holds. Following the lines of the proof of (6.50), we can prove that the controller obtained by (6.47)–(6.54) can guarantee the GUAS of system (6.1) with γ-disturbance attenuation level. The proof is completed. $\square$

6.5 ILLUSTRATIVE EXAMPLES

Example 6.1. Consider the following switched linear system consisting of two subsystems:

$$A_1 = \begin{bmatrix} b & -0.5 \\ a & -0.9 \end{bmatrix}, \quad A_2 = \begin{bmatrix} 0.9 & -a \\ -0.6 & 0.8 \end{bmatrix}.$$

For $\mu = 2$ and $\alpha = 0.1$, Fig. 6.2 presents the stability regions in the plane (a, b). The stability regions of N-step ahead LF approach with the requirements $N > 1$ and $N = 1$ (i.e., the LF approach commonly used under the existing ADT design framework) are drawn to make a simple comparison. It is quite clear that if the number N increases, then the N-step ahead approach yields a larger stability region.

The ADT guaranteeing GUAS for $N = 1$, $N = 2$, $N = 3$, and $N = 4$ can be solved as $\tau_{a1}^* = 6.5788$, $\tau_{a2}^* = 13.1576$, $\tau_{a3}^* = 19.7364$, and $\tau_{a4}^* = 26.3153$, respectively. This implies that when the monotonicity requirements of the LF are relaxed, the switching must be gentler.

For $N = 2$, we choose a pair of $(a, b) = (0, 0.7)$ within the stability region to plot the evolution of LF. As shown in Fig. 6.3, the proposed approach reduces monotonic conservatism by allowing the LF to increase both at switching instants, which are marked by green circles (light gray in print version), and during the subsystem.

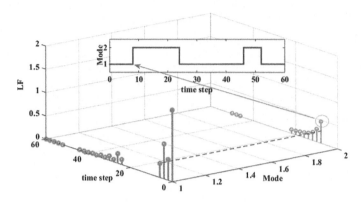

FIGURE 6.3   Evolution of the LF.

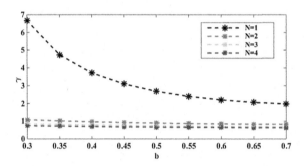

FIGURE 6.4   Comparison of the attenuation level.

Now fixing $a = -0.1$ and setting $B_1 = \begin{bmatrix} 0.3 \\ 0.5 \end{bmatrix}$, $B_2 = \begin{bmatrix} 0.1 \\ 0.7 \end{bmatrix}$, $C_1 = \begin{bmatrix} 0.1 & -0.02 \\ 0.1 & -0.1 \end{bmatrix}$, and $C_2 = \begin{bmatrix} 0.2 & -0.02 \\ 0.1 & -0.1 \end{bmatrix}$, we make a comparison of the l_2-gain γ for $N = 1$, $N = 2$, $N = 3$, and $N = 4$ in Fig. 6.4. Obviously, our approach achieves a better disturbance attenuation performance with increasing the number N.

Example 6.2. We introduce a modified population ecological system model borrowed from [29] as a practical background to illustrate the effectiveness of the H_∞ robust controller presented in Theorem 6.3 for $N = 2$. The differential equations for the population ecological system are as follows:

$$\begin{bmatrix} \dot{Y}_1 \\ \dot{Y}_2 \end{bmatrix} = \begin{bmatrix} \alpha_1^{(\sigma)}(1 - \rho_1^{(\sigma)}) & b_{12}^{(\sigma)} \\ b_{21}^{(\sigma)} & \alpha_2^{(\sigma)}(1 - \rho_2^{(\sigma)}) \end{bmatrix} \begin{bmatrix} Y_1 \\ Y_2 \end{bmatrix} + \begin{bmatrix} c_1^{(\sigma)} \\ c_2^{(\sigma)} \end{bmatrix} u, \quad (6.60)$$

where T_i ($i = 1, 2$) denotes two types of population in the switching environment $\mathbb{E}_\sigma$, $\sigma = 1, 2$; Y_i is the number of individuals, and $\alpha_i^{(\sigma)}$ is the maximum per capita rate of change of T_i, b_{ij}^σ ($j = 1, 2$ and $i \neq j$) is a transfer coefficient modeling the mutual influence of Y_i, and c_i^σ is the effect of immigration ($I_{g,i}$) and emigration ($E_{g,i}$) on Y_i. We denote the carrying capacity of the population by $\lambda_i^{(\sigma)}$, and $\rho_i^{(\sigma)}$ is the approximate proportion of $\lambda_i^{(\sigma)}$ and Y_i. Differently from [29], we consider the difference between the immigration and emigration of T_i as control input to system, i.e., $u = I_{g,i} - E_{g,i}$.

Set the quaternary $(\alpha_i^{(\sigma)}, \rho_i^{(\sigma)}, b_{ij}^{(\sigma)}, c_i^{(\sigma)})$ as $(0.1, 0.5, -0.3, 1.2)$ for $i = 1$, $\sigma = 1$; $(0.4, 2, -0.2, 0)$ for $i = 2$, $\sigma = 1$; $(0.1, 0.5, 0.2, 1)$ for $i = 1$, $\sigma = 2$; and $(0.1, 2, -0.7, 1)$ for $i = 2$, $\sigma = 2$. The discretization model of the selected parameters can be obtained as

$$A_1 = \begin{bmatrix} 1.1 & -0.02 \\ -0.2 & 0.7 \end{bmatrix}, A_2 = \begin{bmatrix} 1 & 0.2 \\ -0.7 & 0.8 \end{bmatrix},$$

$$B_1 = \begin{bmatrix} 1.3 \\ -0.1 \end{bmatrix}, B_2 = \begin{bmatrix} -1.1 \\ 0.6 \end{bmatrix}.$$

Some system parameters and uncertainties are given as

$$B_{1\omega} = \begin{bmatrix} 1 \\ 1 \end{bmatrix}, B_{2\omega} = \begin{bmatrix} 1 \\ 1 \end{bmatrix}, C_1 = \begin{bmatrix} 0.1 & -0.02 \\ 0.1 & -0.1 \end{bmatrix},$$

$$C_2 = \begin{bmatrix} 0.1 & -0.02 \\ 0.1 & -0.1 \end{bmatrix}, D_1 = \begin{bmatrix} 0.3 \\ 0.7 \end{bmatrix}, D_2 = \begin{bmatrix} 0.1 \\ 0.3 \end{bmatrix},$$

$$S_1 = \begin{bmatrix} 0.2 & 0 \\ 0.1 & 0 \end{bmatrix}, S_2 = \begin{bmatrix} 0.1 & 0.1 \\ 0.2 & 0.1 \end{bmatrix},$$

$$H_1 = \begin{bmatrix} 0.1 & 0.2 \\ 0.1 & 0.1 \end{bmatrix}, H_2 = \begin{bmatrix} 0.1 & 0.1 \\ 0.2 & 0.1 \end{bmatrix},$$

$$N_1 = \begin{bmatrix} 0.1 \\ 0.2 \end{bmatrix}, N_2 = \begin{bmatrix} 0.1 \\ 0.2 \end{bmatrix}.$$

We choose $\mu = 2$, $\alpha = 0.1$, the initial state $x_0 = [6000 \ 1000]$, and exogenous disturbance $\omega_k = 5e^{-0.2k}$. The simulation time is taken as 100 time units, and each unit length is taken as $T_s = 2$. By solving condition (6.54) the minimal ADT is obtained as $\tau_a^* = 13.1576$. Then we generate the switching signal path randomly as shown in Fig. 6.5, which satisfies the ADT constraint $\tau_a > \tau_a^*$ with $\tau_a = 15.3846$.

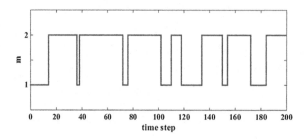

FIGURE 6.5 Switching signals.

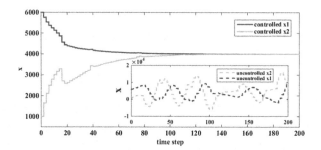

FIGURE 6.6 State trajectories.

The controller gain K_i and optimized attenuation level can be obtained by solving the LMIs given in Theorem 6.3. Therefore we have $K_1 = [\ 0.23\quad -0.03\ ]$, $K_2 = [\ -0.41\quad -0.26\ ]$, and $\gamma = 0.665$. As shown in Fig. 6.6, state trajectories of the system under control and without control are drawn with solid lines and dotted lines, respectively. It is clearly observed that the controlled system is GUAS and state trajectories are regulated to the equilibrium $(4000, 4000)$, which satisfactorily demonstrates that the robust controller design via our approach is effective.

6.6 CONCLUSION

An N-step ahead LF approach has been proposed in this chapter for analysis and synthesis of switched systems with ADT switching. Such a new design approach has effectively reduced the disadvantages of the existing monotonic LF method, and therefore more relaxed stability and l_2-gain results have been achieved. By taking the exogenous disturbance and norm-bounded uncertainties into consideration a robust state feedback controller utilizing the aforementioned approach is further designed and successfully applied to an ecology system. A possible future study could extend the present results to the switched systems with time delay.

References

[1] C. Yuan, F. Wu, Hybrid control for switched linear systems with average dwell time, IEEE Transactions on Automatic Control 60 (1) (2015) 240–245.

[2] D. Liberzon, Switching in Systems and Control, Springer Science & Business Media, 2012.

[3] D. Dai, C.K. Tse, X. Ma, Symbolic analysis of switching systems: application to bifurcation analysis of dc/dc switching converters, IEEE Transactions on Circuits and Systems I: Regular Papers 52 (8) (2005) 1632–1643.

[4] R. Sakthivel, M. Joby, P. Shi, et al., Robust reliable sampled-data control for switched systems with application to flight control, International Journal of Systems Science 47 (15) (2016) 3518–3528.

[5] R. Sakthivel, S. Santra, K. Mathiyalagan, et al., Observer-based control for switched networked control systems with missing data, International Journal of Machine Learning & Cybernetics 6 (4) (2015) 677–686.

[6] B. Niu, C.K. Ahn, H. Li, et al., Adaptive control for stochastic switched nonlower triangular nonlinear systems and its application to a one-link manipulator, IEEE Transactions on Systems, Man, and Cybernetics: Systems (2017) (Early Access).

[7] B. Niu, Y. Liu, G. Zong, et al., Command filter-based adaptive neural tracking controller design for uncertain switched nonlinear output-constrained systems, IEEE Transactions on Cybernetics 47 (10) (2017) 3160–3171.

[8] B. Niu, L. Li, Adaptive backstepping-based neural tracking control for MIMO nonlinear switched systems subject to input delays, IEEE Transactions on Neural Networks and Learning Systems (2017) (Early Access).

[9] X. Zhao, S. Yin, H. Li, et al., Switching stabilization for a class of slowly switched systems, IEEE Transactions on Automatic Control 60 (1) (2015) 221–226.

[10] X. Zhao, Y. Yin, B. Niu, et al., Stabilization for a class of switched nonlinear systems with novel average dwell time switching by T–S fuzzy modeling, IEEE Transactions on Cybernetics 46 (8) (2016) 1952–1957.

[11] X. Zhao, L. Zhang, P. Shi, et al., Stability of switched positive linear systems with average dwell time switching, Automatica 48 (6) (2012) 1132–1137.

[12] P. Zhao, R. Nagamune, Switching LPV controller design under uncertain scheduling parameters, Automatica 76 (2017) 243–250.

[13] M.T. Raza, A.Q. Khan, G. Mustafa, et al., Design of fault detection and isolation filter for switched control systems under asynchronous switching, IEEE Transactions on Control Systems Technology 24 (1) (2016) 13–23.

[14] L. Zhang, P. Shi, l_2–l_∞ model reduction for switched LPV systems with average dwell time, IEEE Transactions on Automatic Control 53 (10) (2008) 2443–2448.

[15] L. Zhang, P. Shi, Stability, l_2-gain and asynchronous, control of discrete-time switched systems with average dwell time, IEEE Transactions on Automatic Control 54 (9) (2009) 2192–2199.

[16] L. Zhang, N. Cui, M. Liu, et al., Asynchronous filtering of discrete-time switched linear systems with average dwell time, IEEE Transactions on Circuits and Systems I: Regular Papers 58 (5) (2011) 1109–1118.

[17] W.A. Zhang, L. Yu, Stability analysis for discrete-time switched time-delay systems, Automatica 45 (10) (2009) 2265–2271.

[18] S.F. Derakhshan, A. Fatehi, Non-monotonic Lyapunov functions for stability analysis and stabilization of discrete time Takagi–Sugeno fuzzy systems, International Journal of Innovative Computing, Information & Control 10 (4) (2014) 1567–1586.

[19] A. Kruszewski, R. Wang, T.M. Guerra, Nonquadratic stabilization conditions for a class of uncertain nonlinear discrete time T-S fuzzy models: a new approach, IEEE Transactions on Automatic Control 53 (2) (2008) 606–611.

[20] A.A. Ahmadi, P.A. Parrilo, Non-monotonic Lyapunov functions for stability of discrete time nonlinear and switched systems, in: Proceedings of the 47th IEEE Conference on Decision and Control, 2008, pp. 614–621.

[21] S.F. Derakhshan, A. Fatehi, M.G. Sharabiany, Nonmonotonic observer-based fuzzy controller designs for discrete time T-S fuzzy systems via LMI, IEEE Transactions on Cybernetics 44 (12) (2014) 2557–2567.

[22] S.F. Derakhshan, A. Fatehi, Non-monotonic robust H_2 fuzzy observer-based control for discrete time nonlinear systems with parametric uncertainties, International Journal of Systems Science 46 (12) (2015) 2134–2149.

[23] A. Kruszewski, T.M. Guerra, New approaches for the stabilization of discrete Takagi–Sugeno fuzzy models, in: Proceedings of the 44th IEEE Conference on Decision and Control and 2005 European Control Conference, 2005, pp. 3255–3260.

[24] T.M. Guerra, A. Kruszewski, M. Bernal, Control law proposition for the stabilization of discrete Takagi–Sugeno models, IEEE Transactions on Fuzzy Systems 17 (3) (2009) 724–731.

[25] Y.J. Chen, M. Tanaka, K. Inoue, et al., A nonmonotonically decreasing relaxation approach of Lyapunov functions to guaranteed cost control for discrete fuzzy systems, IET Control Theory & Applications 8 (16) (2014) 1716–1722.

[26] J. Wen, S.K. Nguang, P. Shi, et al., Robust H_∞ control of discrete-time nonhomogenous Markovian jump systems via multistep Lyapunov function approach, IEEE Transactions on Systems, Man, and Cybernetics: Systems 47 (7) (2017) 1439–1450.

[27] A. Nasiri, S.K. Nguang, A. Swain, et al., Reducing conservatism in H_∞ robust state feedback control design of T-S fuzzy systems: a non-monotonic approach, IEEE Transactions on Fuzzy Systems (2017) (Early Access).

[28] A. Nasiri, S.K. Nguang, A. Swain, et al., Robust output feedback controller design of discrete-time Takagi–Sugeno fuzzy systems: a non-monotonic Lyapunov approach, IET Control Theory & Applications 10 (5) (2016) 545–553.

[29] Y. Ren, M.J. Er, G. Sun, Constraint admissible state sets for switched systems with average dwell time, Automatica 82 (2017) 235–242.

Robust H_∞ Filtering for Average Dwell-Time Switched Systems via a Nonmonotonic Function Approach

7.1 INTRODUCTION

As one of the main problems in control theory [1], filtering problem has attracted great research interests over the past decades due to its wide applications in many fields [2], [3], such as dealing with fault detection [4], [5]. Various filters are designed to reconstruct the unmeasurable states or filter the external noise, and one of the most well-known and effective filters is the Kalman filter [6], which has shown excellent filtering performance by minimizing the variance of the estimate error. However, such an approach become invalid when a priori statistical knowledge of the exogenous noise is not precisely known. To deal with this problem, alternative approaches such as H_2 filtering [7] and H_∞ filtering [8] are introduced to solve the filtering problem with noise with unknown statistics. For example, H_∞ filtering is quite prominent since it allows noises to be arbitrary only with bounded energy or average power level. The aim of the H_∞ filtering is to minimize the H_∞ norm of the filtering error.

Theoretical investigation of H_∞ filtering for switched systems has been a research frontier in control systems in recent years. Reference [9] addressed the H_∞ filtering problem for time-delay [10] switched systems, and then mode-dependent H_∞ filters were further introduced for such systems with nonlinearities by employing the sojourn probability information [11]. When considering the uncertainties, a robust H_∞ filtering problem for uncertain switched systems was addressed in [12].

Generally, a switched system is composed of a finite number of subsystems and a rule orchestrating the switching among them [13]. The average dwell-time (ADT) switching is a typical type of the switching logic, which has been well addressed as a flexible and effective technique for analysis and synthesis of switched systems [14]. Under the framework of ADT switching, the H_∞ filtering for switched systems has been extensively studied. For instance, the exponential H_∞ filtering was developed for uncertain discrete-time switched systems subject to ADT switching [15], which was further extended to switched delay systems in [16]. When asynchronous ADT switching is considered, reference [17] designed mode-dependent filters, where the unmatched H_∞ filters are allowed to perform during the interval of asynchronous switching before applying the matched ones. The results were soon extended to the same systems with time-delay and missing measurements [18].

It is worth noting that due to the mode-independence property of ADT switching law, all subsystems have the same ADT, and a certain conservativeness is introduced. Therefore, a study of the stability and stabilization for switched linear systems with a mode-dependent ADT (MDADT) is launched in [19] from conservatism reduction point of view. Unlike the traditional ADT switching, such a switching strategy is demonstrated to be more general and applicable, where each mode of the underlying system has its own ADT. By utilizing the MDADT technique, a class of discrete-time switched systems consisting of both stable and unstable subsystems can achieve exponential stability under a slow switching scheme and even in the presence of fast switching of unstable subsystems [20]. Apart from stability analysis, MDADT switching was applied in finite-time filtering, weighted H_∞ performance analysis, and finite-time H_∞ control of switched systems to derive less conservative conditions [21], [22], [23]. In this chapter, the MDADT switching method is employed to the study of H_∞ filtering problem for switched systems such that the restrictions on switching can be relaxed to some extent.

Another conservatism of H_∞ filtering for ADT switching systems stems from the monotonic requirement on Lyapunov functions (LFs). For switched systems under ADT switching, the nonmonotonic behaviors of LFs are only allowed at the switching instant. Inspired by the nonmonotonic LF approach [24], which is employed for Takagi–Sugeno (T-S) fuzzy systems to reduce the conservatism [25], this chapter is devoted to developing the nonmonotonic H_∞ filtering approach for switched systems with MDADT switching. Previous works on nonmonotonic LFs are mainly concerned with two-sample variation, i.e., $V(x_{k+2}) < V(x_k)$ [26], N-sample variation, i.e., $V(x_{k+N}) < V(x_k)$ [27], and with a more general case, i.e.,
$$\sum_{i=1}^{N} (V_i(x_{k+i}) - V_i(x_k)) < 0$$ [28]. However, for ADT switching systems, the research of H_∞ filtering based on nonmonotonic LFs still remains open due probably to the mathematical complexity, especially when the N-step ahead scenario [29] is considered.

Motivated by the above discussions, this chapter aims at designing robust H_∞ filters for discrete-time uncertain switched systems subject to MDADT switching via the N-step ahead LF approach. Such an approach allows the LF to increase both at the switching instants and during the running time of each subsystem. The exponential decaying law for the decreasing points of the LF is constructed by determining the joint point between the switching interval $[k_l, k_{l+1}]$ and the predictive horizon $k + N$. The main contributions and novelty of this chapter are summarized as follows:

- A nonmonotonic approach with N-step ahead LF is developed to design a robust H_∞ filter for a discrete-time uncertain switched system such that better filtering performance can be achieved.
- To relax the restriction on the switching law, the MDADT switching is introduced to reduce the ADT bound.

The layout of the remainder is as follows. Section 7.2 briefly describes the system and problem formulation. By employing the N-step ahead LF approach, H_∞ performance analysis is presented in Section 7.3. In Section 7.4, the problem of robust H_∞ filtering is formulated. An example is given in Section 7.5 to demonstrate the potential of the proposed techniques, and conclusions are drawn in Section 7.6.

Notations: Throughout this chapter, $\mathbb{R}^n$ represents the n-dimensional Euclidean space, and $\mathbb{Z}^+$ is the set of nonnegative integers. In addition, the superscript "T" stands for matrix transposition, the symbol $*$ is an ellipsis for the symmetric terms, and $diag\{\cdots\}$ stands for a block-diagonal matrix; $l_2[0, \infty]$ is the space of square-summable infinite sequence, and for $\omega = \{\omega(k)\} \in l_2[0, \infty]$, its norm is given by $||\omega||_2 = \sqrt{\sum_{k=0}^{\infty} |\omega(k)|^2}$. Furthermore, a function $\beta : [0, \infty) \times [0, \infty) \to [0, \infty)$ is said to be of class $\mathcal{KL}$ if $\beta(\cdot, t)$ is of class $\mathcal{K}$ for each fixed $t \geq 0$ and $\beta(s, t)$ decreases to 0 as $t \to \infty$ for each fixed $s \geq 0$. The notation $P > 0$ ($P \geq 0$) means that P is positive definite (semidefinite), and $\aleph_{1,r} \overset{\Delta}{=} \{1, 2, 3, \ldots, r\}$, where $r \in \mathbb{Z}^+$.

7.2 SYSTEM DESCRIPTIONS AND PRELIMINARIES

Consider the following discrete-time switched system:

$$\begin{cases} x(k + 1) = (A_{\sigma(k)} + \Delta A_{\sigma(k)}^\eta)x(k) + B_{\sigma(k)}\omega(k), \\ y(k) = C_{\sigma(k)}x(k) + D_{\sigma(k)}\omega(k), \\ z(k) = L_{\sigma(k)}x(k) + J_{\sigma(k)}\omega(k), \end{cases} \tag{7.1}$$

where $x(k) \in \mathbb{R}^{n_x}$ is the state vector of the system, $y(k) \in \mathbb{R}^{n_y}$ is the regulated output, $\omega(k) \in \mathbb{R}^{n_\omega}$ is the disturbance input, which belongs to

$l_2[0, \infty]$, and $z(k) \in \mathbb{R}^{n_z}$ is the objective signal to be estimated; $\sigma(k)$ is called a switching signal and takes values in the finite set $S = \aleph_{1,r}$, r is the number of subsystems, and $\sigma(k) = i$ means that the ith subsystem is activated. The matrices A_i, B_i, C_i, D_i, L_i, and J_i are system matrices of compatible dimensions. The uncertainties are assumed as $\Delta A_i^\eta = S_i F_\eta H_i$, where S_i and H_i are known constant matrices of appropriate dimensions, and F_η, $\eta \in [k, k + N]$, are unknown time-varying matrix function satisfying $F_\eta^T F_\eta < I$. For convenience, we set $\bar{A}_{\sigma(k)} \triangleq A_{\sigma(k)} + \Delta A_{\sigma(k)}^\eta$.

In this chapter, we are interested in designing a full-order filter for system (7.1) of the following structure:

$$\begin{cases} x_F(k + 1) = A_{Fi} x_F(k) + B_{Fi} y(k), \\ z_F(k) = C_{Fi} x_F(k) + D_{Fi} y(k), \end{cases} \tag{7.2}$$

where A_{Fi}, B_{Fi}, C_{Fi}, and D_{Fi} are filter matrices of appropriate dimensions to be designed. Combining (7.1) with (7.2), the resulting filter error system becomes

$$\begin{cases} \tilde{x}(k + 1) = \tilde{A}_i \tilde{x}(k) + \tilde{E}_i \omega(k), \\ e(k) = \tilde{C}_i \tilde{x}(k) + \tilde{F}_i \omega(k), \end{cases} \tag{7.3}$$

where $\tilde{x}(k) = [\ x^T(k) \quad x_F^T(k)\ ]^T$ is the augmented state vector, $e(k) = z(k) - z_F(k)$ is the filtering error, and

$$\tilde{A}_i = \begin{bmatrix} \bar{A}_i & 0 \\ B_{Fi} C_i & A_{Fi} \end{bmatrix}, \tilde{E}_i = \begin{bmatrix} B_i \\ B_{Fi} D_i \end{bmatrix},$$

$$\tilde{C}_i = [\ L_i - D_{Fi} C_i \quad -C_{Fi}\ ], \tilde{F}_i = J_i - D_{Fi} D_i.$$

Now we have the following preliminaries, which will be essential for further development.

Definition 7.1. System (7.3) with $\omega(k) = 0$ is globally uniformly asymptotically stable (GUAS) under certain switching signal $\sigma(k)$ if for any $k > k_0$, there exists a $\mathcal{KL}$ function β such that the solution $\tilde{x}_k$ of the system satisfies $\|\tilde{x}(k)\| \leq \beta(\|\tilde{x}(k_0)\|, k)$.

Definition 7.2. [17] For a switching signal $\sigma(k)$ and switching time instants $K > k > k_0$, let $N_\sigma(K, k)$ stand for the switching numbers of $\sigma(k)$ over the interval $[k, K)$. If $N_\sigma(K, k) \leq N_0 + (K - k)/\tau_a$ for any given $N_0 > 0$ and $\tau_a > 0$, then τ_a and N_0 are referred to as ADT and the chatter bound, respectively.

Definition 7.3. [19] For a finite switching sequence $\sigma(k)$ and any $K > k > k_0$, let $N_{\sigma i}(K, k)$, $i \in S$, denote the switching number such that ith subsystem are activated over the interval $[k, K)$, and K_i is the total running time of the ith subsystem over the interval $[k, K)$. If there exists $N_{0i} > 0$ (the mode-dependent chatter bounds) such that

$$N_{\sigma i}(K, k) \leq N_{0i} + K_i / \tau_{ai}, \tag{7.4}$$

then we call τ_{ai} the MDADT of the ith subsystem.

Lemma 7.1. *[19] For the discrete-time switched system $x(k+1) = A_{\sigma(k)}x(k)$, let $0 < \alpha_i < 1$ and $\mu_i \geq 1$ be given constants. Assume that there exist C^1 functions $V_{\sigma(k)} : \mathbb{R}^n \to \mathbb{R}$ and two functions κ_1 and κ_2 of class $\mathcal{K}_\infty$ such that, for all $(\sigma(k_l) = i, \sigma(k_l^-) = j) \in S \times S$, we have*

$$\kappa_{1i}(\|x_k\|) \leq V_i(x_k) \leq \kappa_{2i}(\|x_k\|),$$

$$V_i(x_{k+1}) - V_i(x_k) \leq -\alpha_i V_i(x_k),$$

$$V_i(x_{k_l}) \leq \mu_i V_j(k_l^-),$$

where k_l denotes switching instant. Then the system is GUAS with MDADT constraint

$$\tau_{ai} > \tau_{ai}^* = -\ln \mu_i / \ln(1 - \alpha_i). \tag{7.5}$$

Then our main objective is to design a robust H_∞ filter and find admissible MDADT switching signals with MDADT such that filtering error system (7.3) with $\omega(k) = 0$ is GUAS and for given H_∞ norm bound γ [30], the filtering error system (7.3) guarantees

$$\|e\|_2^2 \leq \gamma^2 \|\omega\|_2^2 \tag{7.6}$$

for all nonzero $\omega(k) \in l_2[0, \infty]$ and all admissible uncertainties under zero-initial conditions.

7.3 H_∞ PERFORMANCE ANALYSIS

In this section, without taking uncertainty into consideration, we develop an N-step ahead LF approach to analyze the H_∞ performance of the filtering error system (7.3).

Theorem 7.1. *Consider the filtering error system (7.3) without uncertainties. Let $0 < \alpha_i < 1$ and $\mu_i > 1$ be given constants. Suppose that there exist a scalar $\gamma > 0$, a set of symmetric matrices $Q_{i,f}$ ($f \in \aleph_{1,N-1}$), and $P_i > 0$ such that for $N = 2$,*

$$\begin{bmatrix} -P_i & 0 & P_i \tilde{A}_i & P_i \tilde{E}_i \\ * & -I & \tilde{C}_i & \tilde{F}_i \\ * & * & -Q_{i,1} & 0 \\ * & * & * & -\gamma^2 I \end{bmatrix} \leq 0, \tag{7.7}$$

$$\begin{bmatrix} -Q_{i,1} & 0 & Q_{i,1}\tilde{A}_i & Q_{i,1}\tilde{E}_i \\ * & -I & \tilde{C}_i & \tilde{F}_i \\ * & * & (\alpha_i - 1)P_i & 0 \\ * & * & * & -\gamma^2 I \end{bmatrix} \leq 0, \tag{7.8}$$

for $N \geq 3$ and $f \in \aleph_{2,N-1}$,

$$\begin{bmatrix} -P_i & 0 & P_i\tilde{A}_i & P_i\tilde{E}_i \\ * & -I & \tilde{C}_i & \tilde{F}_i \\ * & * & -Q_{i,N-1} & 0 \\ * & * & * & -\gamma^2 I \end{bmatrix} \leq 0, \tag{7.9}$$

$$\begin{bmatrix} -Q_{i,N-f+1} & 0 & Q_{i,N-f+1}\tilde{A}_i & Q_{i,N-f+1}\tilde{E}_i \\ * & -I & \tilde{C}_i & \tilde{F}_i \\ * & * & -Q_{i,N-f} & 0 \\ * & * & * & -\gamma^2 I \end{bmatrix} \leq 0, \tag{7.10}$$

$$\begin{bmatrix} -Q_{i,1} & 0 & Q_{i,1}\tilde{A}_i & Q_{i,1}\tilde{E}_i \\ * & -I & \tilde{C}_i & \tilde{F}_i \\ * & * & (\alpha_i - 1)P_i & 0 \\ * & * & * & -\gamma^2 I \end{bmatrix} \leq 0, \tag{7.11}$$

for all $N \in \mathbb{Z}^+$,

$$P_i \leq \mu_i P_j, \tag{7.12}$$

for all $(\sigma(k_l) = i, \sigma(k_l^-) = j) \in S \times S$, and the MDADT of the switching signal $\sigma(k)$ satisfies

$$\tau_{ai} > \tau_{ai}^* = -N \ln \mu_i / \ln(1 - \alpha_i). \tag{7.13}$$

Then system (7.3) is GUAS with the H_∞ norm bound γ.

Proof. Before performing H_∞ performance analysis, we first address sufficient conditions guaranteeing GUAS of system (7.3) without considering exogenous disturbance and uncertainties.

Consider the quadratic LF

$$V_i(\tilde{x}(k)) = \tilde{x}^T(k)P_i\tilde{x}(k). \tag{7.14}$$

Taking the N-step time difference of this LF,

$$\Delta V_{i,N}(\tilde{x}(k)) = V_i(\tilde{x}(k+N)) - V_i(\tilde{x}(k)), \tag{7.15}$$

consider the auxiliary function

$$V_i'(\tilde{x}(k+f)) = \tilde{x}^T(k+f)Q_{i,f}\tilde{x}(k+f), \quad f \in \aleph_{1,N-1}.$$

If conditions (7.9)–(7.11) hold for $N \geq 3$, then by some basic matrix manipulations we can obtain

$$\begin{bmatrix} -P_i & P_i \tilde{A}_i \\ * & -Q_{i,N-1} \end{bmatrix} \leq 0, \tag{7.16}$$

$$\begin{bmatrix} -Q_{i,N-f+1} & Q_{i,N-f+1} \tilde{A}_i \\ * & -Q_{i,N-f} \end{bmatrix} \leq 0, \ f \in \aleph_{2.N-1}, \tag{7.17}$$

$$\begin{bmatrix} -Q_{i,1} & Q_{i,1} \tilde{A}_i \\ * & (\alpha_i - 1) P_i \end{bmatrix} \leq 0. \tag{7.18}$$

Assuming the zero disturbance input to the system, from (7.16)–(7.18) we can infer the following inequalities along the trajectory of system (7.3):

$$V_i(\tilde{x}(k+N)) - V_i'(\tilde{x}(k+N-1)) \leq 0, \tag{7.19}$$

$$V_i'(\tilde{x}(k+N-f+1)) - V_i'(\tilde{x}(k+N-f)) \leq 0, \tag{7.20}$$

$$V_i'(\tilde{x}(k+1)) + (\alpha_i - 1) V_i(\tilde{x}(k)) \leq 0. \tag{7.21}$$

Therefore, the equation $V_i(\tilde{x}(k+N)) + (\alpha_i - 1) \ V_i(\tilde{x}(k)) \leq 0$ holds, i.e.,

$$V_i(\tilde{x}(k+N)) - V_i(\tilde{x}(k)) \leq -\alpha_i V_i(\tilde{x}(k)). \tag{7.22}$$

Following a similar routine to the above derivation, from conditions (7.7)–(7.8) we can obtain that (7.22) also holds for $N = 2$.

Under the assumption that the switching behavior is much slower than the sampling time step, it follows from (7.22) that

$$V_i(\tilde{x}(k)) \leq (1 - \alpha_i)^{\frac{k-k_l}{N}} V_i(\tilde{x}(k_l)). \tag{7.23}$$

From (7.12) it follows that

$$V_i(\tilde{x}(k_l)) \leq \mu_i V_j(\tilde{x}(k_l^-)), \tag{7.24}$$

where $k_l, k_{l-1}, \ldots, k_1, k_0$ denote the switching times, and $\sigma(k) = \sigma(k_l) = i$, $\sigma(k_l^-) = \sigma(k_{l-1}) = j$. Combining (7.23) and (7.24) and setting $(1 - \alpha_{\sigma(k)}) = \bar{\alpha}_{\sigma(k)}$, we readily obtain

$$V_{\sigma(k)}(\tilde{x}(k)) \leq \bar{\alpha}_{\sigma(k)}^{\frac{k-k_l}{N}} V_{\sigma(k)}(\tilde{x}(k_l))$$

$$\leq \mu_{\sigma(k)} \bar{\alpha}_{\sigma(k)}^{\frac{k-(k_l^-)}{N}} V_{\sigma(k_l^-)}(\tilde{x}((k_l^-)))$$

$$\leq \mu_{\sigma(k)} \bar{\alpha}_{\sigma(k)}^{\frac{k-k_l}{N}} \bar{\alpha}_{\sigma(k_{l-1})}^{\frac{k-k_{l-1}}{N}} V_{\sigma(k_{l-1})}(\tilde{x}(k_{l-1}))$$

$$\vdots$$

$$\leq \prod_{s=1}^{N_\sigma(k,k_0)} \mu_{\sigma(k_s)} \bar{\alpha}_{\sigma(k)}^{\frac{k-k_l}{N}} \bar{\alpha}_{\sigma(k_{l-1})}^{\frac{k-k_{l-1}}{N}} \cdots \bar{\alpha}_{\sigma(k_1)}^{\frac{k_2-k_1}{N}} \bar{\alpha}_{\sigma(k_0)}^{\frac{k_1-k_0}{N}} V_{\sigma(k_0)}(\tilde{x}(k_0)).$$

Assume that the switching signal subjects to the set $\{1, 2, \ldots, \Gamma\}$ and $K_p(k, k_0)$ is the total running time of the pth subsystem over the interval $[k_0, k]$. Setting $V_{\sigma(k)}(\tilde{x}(k)) \triangleq V_k$ for convenience, according to (7.4), we obtain

$$
\begin{aligned}
V_k &\le \prod_{p=1}^{\Gamma} (\mu_p^{N_{\sigma p}} \bar{\alpha}_p^{\frac{K_p(k,k_0)}{N}}) V_{k_0} \\
&\le \prod_{p=1}^{\Gamma} (\mu_p^{N_{0p}+K_p(k,k_0)/\tau_{ap}} \bar{\alpha}_p^{\frac{K_p(k,k_0)}{N}}) V_{k_0} \\
&\le \exp\{\prod_{p=1}^{\Gamma} N_{0p} \ln \mu_p + (k-k_0)[\ln \mu_p^{\frac{1}{\tau_{ap}}} + \ln \bar{\alpha}_p^{\frac{1}{N}}]\} V_{k_0} \\
&\le \exp\{\prod_{p=1}^{\Gamma} N_{0p} \ln \mu_p + (k-k_0)[\ln(\mu_p^{\frac{1}{\tau_{ap}}} \bar{\alpha}_p^{\frac{1}{N}})]\} V_{k_0}.
\end{aligned}
$$

If the MDADT satisfies (7.13), it is straightforward to get $V_k \le \exp\{\prod_{p=1}^{\Gamma} N_{0p} \ln \mu_p\} V_{k_0}$. Therefore, the GUAS of filtering error system (7.3) without disturbance and uncertainties can be inferred by referring to Definition 7.1.

Next, we focus on the H_∞ performance analysis and assume that system (7.3) is under the zero initial condition. Define the H_∞ performance index

$$
J_M \triangleq \| e \|_2^2 - \gamma^2 \| \omega \|_2^2 = \sum_{k=0}^{M-1} \{e_k^T e_k - \gamma^2 \omega_k^T \omega_k\}.
$$

Then we introduce a set of auxiliary matrices such that H_∞ performance analysis can be achieved by moving the horizon from $k + N - 1$ to k step by step. For convenience, we set $\tilde{x}(k) \triangleq \tilde{x}_k$.

Step 1: When $N \ge 3$, adding and subtracting $V_i'(\tilde{x}_{k+N-1}) + e_{k+N-1}^T e_{k+N-1} - \gamma^2 \omega_{k+N-1}^T \omega_{k+N-1}$ to and from (7.15), we can readily obtain

$$
\begin{aligned}
\Delta V_{i,N}(\tilde{x}_k) = V_i(\tilde{x}_{k+N}) &+ V_i'(\tilde{x}_{k+N-1}) - V_i'(\tilde{x}_{k+N-1}) \\
&- (e_{k+N-1}^T e_{k+N-1} - \gamma^2 \omega_{k+N-1}^T \omega_{k+N-1}) \\
&+ e_{k+N-1}^T e_{k+N-1} - \gamma^2 \omega_{k+N-1}^T \omega_{k+N-1} - V_i(\tilde{x}_k).
\end{aligned}
$$

Along the trajectory of system (7.3), we get

$$
\begin{aligned}
\Delta V_{i,N}(\tilde{x}_k) = \tilde{x}_{k+N-1}^T Q_{i,N-1} \tilde{x}_{k+N-1} &- \tilde{x}_k^T P_i \tilde{x}_k \\
&+ \tilde{x}_{k+N-1}^T (\tilde{A}_i^T P_i \tilde{A}_i - Q_{i,N-1} + \tilde{C}_i^T \tilde{C}_i) \tilde{x}_{k+N-1}
\end{aligned}
$$

$$+ 2\tilde{x}_{k+N-1}^T (\tilde{A}_i^T P_i \tilde{E}_i + \tilde{C}_i^T \tilde{F}_i) \omega_{k+N-1}$$
$$+ \omega_{k+N-1}^T (\tilde{F}_i^T P_i \tilde{F}_i - \gamma^2 I + \tilde{E}_i^T \tilde{E}_i) \omega_{k+N-1}$$
$$- (e_{k+N-1}^T e_{k+N-1} - \gamma^2 \omega_{k+N-1}^T \omega_{k+N-1}).$$

From condition (7.9) it follows that

$$\Delta V_{i,N}(\tilde{x}_k) \le \tilde{x}_{k+N-1}^T Q_{i,N-1} \tilde{x}_{k+N-1} - \tilde{x}_k^T P_i \tilde{x} \qquad (7.25)$$
$$- (e_{k+N-1}^T e_{k+N-1} - \gamma^2 \omega_{k+N-1}^T \omega_{k+N-1}).$$

Step 2: Then, adding and subtracting $V_i'(\tilde{x}_{k+N-2}) + e_{k+N-2}^T e_{k+N-2} - \gamma^2 \omega_{k+N-2}^T \omega_{k+N-2}$ to and from (7.25), since (7.10) holds for $f = 2$, we get

$$\Delta V_{i,N}(\tilde{x}_k) \le \tilde{x}_{k+N-2}^T Q_{i,N-2} \tilde{x}_{k+N-2} - \tilde{x}_k^T P_i \tilde{x}_k$$
$$- \sum_{f=N-2}^{N-1} (e_{k+f}^T e_{k+f} - \gamma^2 \omega_{k+f}^T \omega_{k+f}).$$

Repeating the procedure $N - 3$ times yields

$$\Delta V_{i,N}(\tilde{x}_k) \le \tilde{x}_{k+1}^T Q_{i,1} \tilde{x}_{k+1} - \tilde{x}_k^T P_i \tilde{x}_k$$
$$- \sum_{f=1}^{N-1} (e_{k+f}^T e_{k+f} - \gamma^2 \omega_{k+f}^T \omega_{k+f}).$$

Step 3: Adding and subtracting $\alpha_i V_i(\tilde{x}_k) + e_k^T e_k - \gamma^2 \omega_k^T \omega_k$, by (7.11) we get

$$\sum_{f=0}^{N-1} (e_{k+f}^T e_{k+f} - \gamma^2 \omega_{k+f}^T \omega_{k+f}) + \Delta V_{i,N}(\tilde{x}_k) \le -\alpha_i V_i(\tilde{x}_k). \qquad (7.26)$$

Following a similar routine as before, we obtain (7.26) for $N = 2$. Taking the summation from $k = -N + 1$ to $k = M$ on both sides of (7.26), we get

$$\sum_{k=-N+1}^{M} \{\sum_{f=0}^{N-1} (e_{k+f}^T e_{k+f} - \gamma^2 \omega_{k+f}^T \omega_{k+f})\} + \sum_{k=-N+1}^{M} \{\Delta V_{i,N}(\tilde{x}_k)\}$$
$$\le \sum_{k=-N+1}^{M} \{-\alpha_i V_i(\tilde{x}_k)\}.$$

Then, under the zero initial conditions $x_0 = x_{-1} = x_{-2} \cdots = x_{-N+1} = 0$, we have

$$\sum_{k=-N+1}^{M} \{\Delta V_{i,N}(\tilde{x}_k)\} = V_i(\tilde{x}_{M+1}) + V_i(\tilde{x}_{M+2}) + \cdots + V_i(\tilde{x}_{M+N}).$$

As $M \to \infty$, it is straightforward to obtain

$$N V_i(\tilde{x}_\infty) + N J_\infty \le \sum_{k=1}^{\infty} \{-\alpha_i V_i(\tilde{x}_k)\}.$$

It is clear that $J_M < 0$, i.e., $\|e\|_2^2 < \gamma^2 \|\omega\|_2^2$ since $N V_i(\tilde{x}_\infty) \ge 0$ and $\sum_{k=1}^{\infty} \{-\alpha_i V_i(\tilde{x}_k)\} < 0$, and thus we obtain a prescribed disturbance attenuation level γ. The proof is completed.   □

Remark 7.1. Differently from the ADT approach, which neglects the individual properties of each subsystem, the MDADT switching method determines τ_{ai} as the infimum of the average time among the intervals associated with the ith subsystem, and we can conclude that $\tau_{ai}^* \le \tau_a^*$, which means that the restriction on the switching frequency can be relaxed over interval $[k, K)$.

Remark 7.2. It should be noted that $N = 1$ is a particular case of the N-step approach, and the LF in this case is required to decrease monotonically within the subsystems. The stability analysis for $N = 1$ has been addressed in [19], and we omit the H_∞ performance analysis since the corresponding derivation is not tough.

7.4 ROBUST H_∞ FILTERING DESIGN

In this section, we formulate a sufficient condition for the existence of robust H_∞ filter for uncertain filtering error system (7.3).

Theorem 7.2. *Let $0 < \alpha_i < 1$ and $\mu_i > 1$ be given constants. Suppose that there exist scalars $\vartheta_i > 0$ and $\gamma > 0$, matrices $P_{1i} > 0$ and $P_{3i} > 0$, and matrices P_{2i}, X_i, Y_i, Z_i, A_{fi}, B_{fi}, C_{fi}, D_{fi}, $Q_{1i,f}$, $Q_{2i,f}$, $Q_{3i,f}$ ($f \in \aleph_{1,N-1}$) such that, for $N = 2$,*

$$\begin{bmatrix} \Phi_i^{11} & 0 & \Phi_i^{13} & \Phi_i^{14} & \Phi_i^{15} \\ * & -I & \Phi_i^{23} & \Phi_i^{24} & 0 \\ * & * & \Upsilon_i^{33} & 0 & 0 \\ * & * & * & -\gamma^2 I & 0 \\ * & * & * & * & -\vartheta_i I \end{bmatrix} \le 0, \tag{7.27}$$

$$
\begin{bmatrix}
\Upsilon_i^{11} & 0 & \Phi_i^{13} & \Phi_i^{14} & \Phi_i^{15} \\
* & -I & \Phi_i^{23} & \Phi_i^{24} & 0 \\
* & * & \Phi_i^{33} & 0 & 0 \\
* & * & * & -\gamma^2 I & 0 \\
* & * & * & * & -\vartheta_i I
\end{bmatrix} \le 0,
\tag{7.28}
$$

for $N \ge 3$ and $f \in \aleph_{2.N-1}$,

$$
\begin{bmatrix}
\Phi_i^{11} & 0 & \Phi_i^{13} & \Phi_i^{14} & \Phi_i^{15} \\
* & -I & \Phi_i^{23} & \Phi_i^{24} & 0 \\
* & * & \Theta_i^{33} & 0 & 0 \\
* & * & * & -\gamma^2 I 0 & \\
* & * & * & * & -\vartheta_i I
\end{bmatrix} \le 0,
\tag{7.29}
$$

$$
\begin{bmatrix}
\Xi_i^{11} & 0 & \Phi_i^{13} & \Phi_i^{14} & \Phi_i^{15} \\
* & -I & \Phi_i^{23} & \Phi_i^{24} & 0 \\
* & * & \Xi_i^{33} & 0 & 0 \\
* & * & * & -\gamma^2 I & 0 \\
* & * & * & * & -\vartheta_i I
\end{bmatrix} \le 0,
\tag{7.30}
$$

$$
\begin{bmatrix}
\Upsilon_i^{11} & 0 & \Phi_i^{13} & \Phi_i^{14} & \Phi_i^{15} \\
* & -I & \Phi_i^{23} & \Phi_i^{24} 0 & \\
* & * & \Phi_i^{33} & 0 & 0 \\
* & * & * & -\gamma^2 I & 0 \\
* & * & * & * & -\vartheta_i I
\end{bmatrix} \le 0,
\tag{7.31}
$$

for all $N \in \mathbb{Z}^+$, and

$$
\begin{bmatrix}
P_{1i} & P_{2i} \\
* & P_{3i}
\end{bmatrix} - \mu_i
\begin{bmatrix}
P_{1j} & P_{2j} \\
* & P_{3j}
\end{bmatrix} \le 0
\tag{7.32}
$$

for all $(\sigma(k_l) = i, \sigma(k_l^-) = j) \in S \times S$, where

$$
\Phi_i^{11} =
\begin{bmatrix}
P_{1i} - X_i - X_i^T & P_{2i} - Y_i - Z_i^T \\
* & P_{3i} - Y_i - Y_i^T
\end{bmatrix},
$$

$$
\Phi_i^{13} =
\begin{bmatrix}
X_i A_i + B_{fi} & A_{fi} \\
Z_i A_i + B_{fi} & A_{fi}
\end{bmatrix},
\Phi_i^{14} =
\begin{bmatrix}
X_i B_i + B_{fi} D_i \\
Z_i B_i + B_{fi} D_i
\end{bmatrix},
$$

$$\Phi_i^{15} = \begin{bmatrix} X_i S_i & 0 \\ Z_i S_i & 0 \end{bmatrix}, \Phi_i^{23} = \begin{bmatrix} L_i - D_{fi} C_i & -C_{fi} \end{bmatrix},$$

$$\Phi_i^{33} = \begin{bmatrix} (\alpha_i - 1) P_{1i} + \vartheta_i H_i^T H_i & (\alpha_i - 1) P_{2i} \\ * & (\alpha_i - 1) P_{3i} \end{bmatrix},$$

$$\Phi_i^{24} = J_i - D_{fi} D_i, \Upsilon_i^{33} = \begin{bmatrix} -Q_{1i,1} + \vartheta_i H_i^T H_i & -Q_{2i,1} \\ * & -Q_{3i,1} \end{bmatrix},$$

$$\Upsilon_i^{11} = \begin{bmatrix} Q_{1i,1} - X_i - X_i^T & Q_{2i,1} - Y_i - Z_i^T \\ * & Q_{3i,1} - Y_i - Y_i^T \end{bmatrix},$$

$$\Theta_i^{33} = \begin{bmatrix} -Q_{1i,N-1} + \vartheta_i H_i^T H_i & -Q_{2i,N-1} \\ * & -Q_{3i,N-1} \end{bmatrix},$$

$$\Xi_i^{11} = \begin{bmatrix} Q_{1i,N-f+1} - X_i - X_i^T & Q_{2i,N-f+1} - Y_i - Z_i^T \\ * & Q_{3i,N-f+1} - Y_i - Y_i^T \end{bmatrix},$$

$$\Xi_i^{33} = \begin{bmatrix} -Q_{1i,N-f} + \vartheta_i H_i^T H_i & -Q_{2i,N-f} \\ * & -Q_{3i,N-f} \end{bmatrix}.$$

Then the filtering error system (7.3) is GUAS for any switching signal with MDADT satisfying (7.13) and has a prescribed H_∞ noise-attenuation level γ. The filter gains can be obtained as

$$A_{Fi} = Y_i^{-1} A_{fi}, B_{Fi} = Y_i^{-1} B_{fi}, C_{Fi} = C_{fi}, D_{Fi} = D_{fi}.$$

Proof. Pre- and postmultiply (7.9) by $diag\{G_i P_i^{-1}, I, I, I\}$ and $diag\{P_i^{-1} G_i^T, I, I, I\}$, respectively. Then since $-G_i^T - G_i + P_i \leq -G_i P_i^{-1} G_i^T$, (7.9) is transformed into

$$\Psi_1 = \begin{bmatrix} P_i - G_i^T - G_i & 0 & G_i \tilde{A}_i & G_i \tilde{E}_i \\ * & -I & \tilde{C}_i & \tilde{F}_i \\ * & * & -Q_{i,N-1} & 0 \\ * & * & * & -\gamma^2 I \end{bmatrix} \leq 0. \qquad (7.33)$$

Denote

$$\tilde{A}_i = \tilde{A}_i^1 + \tilde{A}_i^2 = \begin{bmatrix} A_i & 0 \\ B_{Fi} C_i & A_{Fi} \end{bmatrix} + \begin{bmatrix} S_i F_\eta H_i & 0 \\ 0 & 0 \end{bmatrix},$$

$$\bar{S}_i = \begin{bmatrix} S_i & 0 \\ 0 & 0 \end{bmatrix}, \bar{F}_\eta = \begin{bmatrix} F_\eta & 0 \\ 0 & 0 \end{bmatrix}, \bar{H}_i = \begin{bmatrix} H_i & 0 \\ 0 & 0 \end{bmatrix},$$

$$\tilde{S}_i = \begin{bmatrix} 0 & 0 & G_i \bar{S}_i & 0 \\ 0 & 0 & 0 & 0 \\ 0 & 0 & 0 & 0 \\ 0 & 0 & 0 & 0 \end{bmatrix}, \tilde{F}_\eta = \begin{bmatrix} 0 & 0 & 0 & 0 \\ 0 & 0 & 0 & 0 \\ 0 & 0 & \bar{F}_\eta & 0 \\ 0 & 0 & 0 & 0 \end{bmatrix}, \tilde{H}_i = \begin{bmatrix} 0 & 0 & 0 & 0 \\ 0 & 0 & 0 & 0 \\ 0 & 0 & \bar{H}_i & 0 \\ 0 & 0 & 0 & 0 \end{bmatrix}.$$

Subsequently, (7.33) can be decomposed as

$$\Psi_1 = \Psi_0 + \tilde{S}_i \tilde{F}_\eta \tilde{H}_i + \left(\tilde{S}_i \tilde{F}_\eta \tilde{H}_i \right)^T,$$

where

$$\Psi_0 = \begin{bmatrix} P_i - G_i^T - G_i & 0 & G_i \tilde{A}_i^1 & G_i \bar{E}_i \\ * & -I & \tilde{C}_i & \bar{F}_i \\ * & * & -Q_{i,N-1} & 0 \\ * & * & * & -\gamma^2 I \end{bmatrix}.$$

By applying the inequality

$$\Psi_0 + \tilde{S}_i \tilde{F}_\eta \tilde{H}_i + \left(\tilde{S}_i \tilde{F}_\eta \tilde{H}_i \right)^T \leq \Psi_0 + \varepsilon_i \tilde{S}_i \tilde{S}_i^T + \varepsilon_i^{-1} \tilde{H}_i^T \tilde{H}_i$$

it follows that

$$\begin{bmatrix} P_i - G_i^T - G_i & * & G_i \tilde{A}_i^1 & G_i \bar{E}_i & G_i \bar{S}_i \\ * & -I & \tilde{C}_i & \bar{F}_i & 0 \\ * & * & -Q_{i,N-1} + \varepsilon_i^{-1} \tilde{H}_i^T \tilde{H}_i & 0 & 0 \\ * & * & * & -\gamma^2 I & 0 \\ * & * & * & * & -\vartheta_i I \end{bmatrix} \leq 0, \tag{7.34}$$

where $\vartheta_i = \varepsilon_i^{-1}$.

Then, replace $\tilde{A}_i$, $\tilde{E}_i$, $\tilde{C}_i$, $\tilde{F}_i$ in (7.34) by those in (7.3) and assume that P_i, G_i, $Q_{i,f}$ have the following forms:

$$P_i \triangleq \begin{bmatrix} P_{1i} & P_{2i} \\ * & P_{3i} \end{bmatrix}, G_i \triangleq \begin{bmatrix} X_i & Y_i \\ Z_i & Y_i \end{bmatrix}, Q_{i,f} \triangleq \begin{bmatrix} Q_{1i,f} & Q_{2i,f} \\ * & Q_{3i,f} \end{bmatrix}.$$

Defining the matrix variables

$$A_{fi} = Y_i A_{Fi}, \; B_{fi} = Y_i B_{Fi}, \; C_{fi} = C_{Fi}, \; D_{fi} = D_{Fi},$$

we can directly obtain that condition (7.29) holds. Similarly, (7.30)–(7.31) and conditions (7.27)–(7.28) for $N = 2$ can be easily derived from Theorem 7.1. Therefore, we can conclude that the filtering error system (7.3) is GUAS with γ-disturbance attenuation level. The proof is completed. $\quad\square$

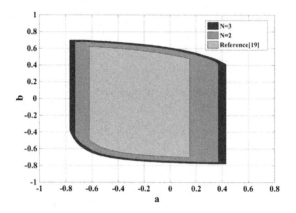

FIGURE 7.1 Comparison of the feasible region for the filtering error system.

7.5 ILLUSTRATIVE EXAMPLES

Example 7.1. Consider a switched system consisting of two subsystems as follows:

$$A_1 = \begin{bmatrix} 0.3 & 0.3 \\ 0.5 & a \end{bmatrix}, A_2 = \begin{bmatrix} a & 0.5 \\ 0.2 & b \end{bmatrix},$$

$$B_1 = \begin{bmatrix} 0.4 \\ -1 \end{bmatrix}, B_2 = \begin{bmatrix} 0.5 \\ 1.2 \end{bmatrix},$$

$$C_1 = \begin{bmatrix} 0.1 & -0.2 \end{bmatrix}, C_2 = \begin{bmatrix} 0.2 & -0.2 \end{bmatrix},$$

$$D_1 = 0.3, \ D_2 = 0.1, \ J_1 = 1.2, \ J_2 = 0.8,$$

$$L_1 = \begin{bmatrix} 1.2 & 0.8 \end{bmatrix}, L_2 = \begin{bmatrix} 0.8 & 0.5 \end{bmatrix},$$

$$S_1 = \begin{bmatrix} 0.2 & 0.3 \\ 0.1 & 0.2 \end{bmatrix}, S_2 = \begin{bmatrix} 0.1 & 0.4 \\ 0.2 & 0.2 \end{bmatrix},$$

$$H_1 = \begin{bmatrix} 0.1 & 0.2 \\ 0.3 & 0.1 \end{bmatrix}, H_2 = \begin{bmatrix} 0.1 & 0.1 \\ 0.2 & 0.5 \end{bmatrix}.$$

The objective of this simulation is to demonstrate the effectiveness of the robust H_∞ filter and find out admissible switching signals such that the filtering performance and switching frequency can be improved.

Firstly, by assigning $\mu_1 = \mu_2 = 2, \alpha_1 = 0.4$, and $\alpha_2 = 0.2$, Fig. 7.1 presents the feasible regions for the filtering error system by employing the conventional approach in [19] (equivalent to the case of $N = 1$) and the proposed

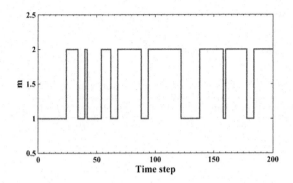

FIGURE 7.2 Switching signals.

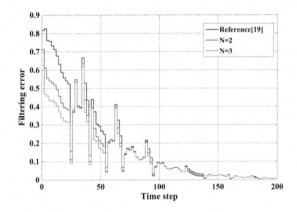

FIGURE 7.3 Comparison of filtering error curves.

approach ($N = 2$ and $N = 3$) in the plane (a, b). It is clear that the proposed approach achieves a larger feasible region.

We choose $a = -0.25$ and $b = -0.2$ within the feasible region and set the initial state $x_0 = x_{F0} = [\ 0 \quad 0\ ]^T$ and exogenous disturbance $\omega_k = 0.5e^{-0.05k}$. The simulation time is taken as 100 time units, and each unit length is taken as $T_s = 2$. Then we generate the switching signal path randomly as shown in Fig. 7.2, which satisfies the minimal MDADT constraints obtained from (7.13). The filtering error response of system (7.3) by employing the conventional approach in [19] and nonmonotonic approach ($N = 2$ and $N = 3$) are drawn in Fig. 7.3. We observe that our approach achieves a better filtering performance with increasing the number N.

Subsequently, we make a comparison of the H_∞ performance index γ for reference [19] and the N-step LF approach addressed in this chapter ($N = 2$ and $N = 3$) in Table 7.1. Obviously, our approach improves the ca-

TABLE 7.1 Comparison of the attenuation level.

Approach	Reference [19]	$N = 2$	$N = 3$
index γ	$\gamma = 3.960$	$\gamma = 2.8645$	$\gamma = 2.642$

TABLE 7.2 Comparison of two different switching schemes.

Scheme	Reference [19] (ADT)	This chapter (MDADT)
$N = 1$	$\tau_{a1}^* = 3.1063$	$\tau_{a11}^* = 1.3569,\ \tau_{a12}^* = 3.1063$
$N = 2$	$\tau_{a2}^* = 6.2126$	$\tau_{a21}^* = 2.7138,\ \tau_{a22}^* = 6.2126$
$N = 3$	$\tau_{a3}^* = 9.3189$	$\tau_{a31}^* = 4.0707,\ \tau_{a32}^* = 9.3189$

pability of disturbance attenuation, and with increasing the number N, a better optimized H_∞ performance can be guaranteed.

Then, to show the advantages of MDADT switching, we make a simple comparison with the ADT switching. We set $\mu = \mu_1 = \mu_2 = 2$, $\alpha = 0.2$, $\alpha_1 = 0.4$, and $\alpha_2 = 0.2$, and the comparison results are listed in Table 7.2. It is quite clear that 1) with increasing the number N, switching is required to be gentler and 2) in contrast to ADT switching, MDADT switching reduces the ADT bound, which means that the restriction on the switching law is relaxed.

Example 7.2. To demonstrate the practical potential of the robust H_∞ filter presented in Theorem 7.2, we introduce a PWM (Pulse-Width-Modulation)-driven boost converter system model borrowed from [17]. The differential equations for the boost converter are as follows:

$$\dot{e}_C(\tau) = -\frac{1}{RC_1}e_C(\tau) + (1 - s(\tau))\frac{1}{C_1}i_L(\tau), \tag{7.35}$$

$$\dot{i}_L(\tau) = -(1 - s(\tau))\frac{1}{L_1}e_C(\tau) + s(\tau)\frac{1}{L_1}e_s(\tau), \tag{7.36}$$

where L denotes the inductance, C is the capacitance, R is the load resistance, $e_S(t)$ is the source voltage, T represents the period, and $\tau = t/T$, $C_1 = C/T$. Then (7.35)–(7.36) can be further expressed as

$$\dot{x} = A_\sigma^c x, \sigma \in \{1, 2\},$$

where $x = [\ e_C \quad i_L \quad 1\ ]^T$, and

$$A_1^C = \begin{bmatrix} -\frac{1}{RC_1} & \frac{1}{C_1} & 0 \\ -\frac{1}{L_1} & 0 & 0 \\ 0 & 0 & 0 \end{bmatrix}, A_2^C = \begin{bmatrix} -\frac{1}{RC_1} & 0 & 0 \\ 0 & 0 & \frac{1}{L_1} \\ 0 & 0 & 0 \end{bmatrix}.$$

According to [17], a set of discretization system matrices is described as follows:

$$A_1 = \begin{bmatrix} 0.94 & 0.10 & 0.06 \\ -0.30 & 0.95 & -0.30 \\ -0.25 & -0.06 & 0.63 \end{bmatrix},$$

$$A_2 = \begin{bmatrix} 0.93 & 0.08 & 0.07 \\ -0.14 & 0.66 & -0.20 \\ -0.16 & -0.40 & 0.66 \end{bmatrix},$$

$$B_1 = \begin{bmatrix} -0.30 \\ 0.20 \\ 0.10 \end{bmatrix}, \quad B_2 = \begin{bmatrix} -1.40 \\ -0.30 \\ 0.20 \end{bmatrix},$$

$$C_1 = [\ 0.10 \quad -0.10 \quad 0.10\ ],$$

$$C_2 = [\ 0.30 \quad -0.40 \quad 0.10\ ],$$

$$D_1 = 0.4, \ D_2 = -0.5, \ L_1 = L_2 = 0,$$

$$H_1 = [\ 0.70 \quad 0 \quad 0.30\ ], H_2 = [\ 0.20 \quad 0 \quad 0.40\ ].$$

When considering the uncertainties, we set

$$S_1 = \begin{bmatrix} 0.1 & 0.1 & 0.1 \\ 0.1 & 0.1 & 0.1 \\ 0.1 & 0.1 & 0.1 \end{bmatrix}, S_2 = \begin{bmatrix} 0.1 & 0.1 & 0.1 \\ 0.2 & 0.2 & 0.1 \\ 0.1 & 0.2 & 0.1 \end{bmatrix},$$

$$H_1 = \begin{bmatrix} 0.1 & 0.2 & 0.1 \\ 0.1 & 0.1 & 0.1 \\ 0.1 & 0.2 & 0.1 \end{bmatrix}, H_2 = \begin{bmatrix} 0.1 & 01 & 0.1 \\ 0.2 & 0.1 & 0.1 \\ 0.1 & 0.2 & 0.1 \end{bmatrix}.$$

By giving $\alpha_1 = 0.02$, $\alpha_2 = 0.01$, and $\mu = 1.02$, the optimized attenuation level can be obtained by solving the conditions given for $N = 2$ in Theorem 7.2. Therefore we have $\gamma = 2.8168$, $\tau_{a1}^* = 1.9604$, and $\tau_{a2}^* = 3.9407$, and the filter gains can be obtained as

$$A_{F1} = \begin{bmatrix} 0.64 & 0.33 & -0.09 \\ -0.01 & 0.59 & -0.02 \\ -0.37 & -0.02 & 0.39 \end{bmatrix},$$

$$A_{F2} = \begin{bmatrix} 0.42 & 0.81 & -0.22 \\ 0.03 & 0.09 & 0.04 \\ -0.13 & -0.27 & 0.37 \end{bmatrix},$$

$$B_{F1} = \begin{bmatrix} -2.38 \\ 3.62 \\ -0.11 \end{bmatrix}, B_{F2} = \begin{bmatrix} -1.05 \\ 0.67 \\ -1.02 \end{bmatrix},$$

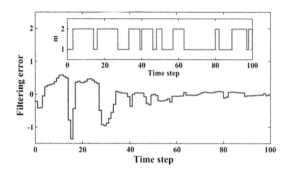

FIGURE 7.4   Filtering error curve for $N = 2$.

$$C_{F1} = \begin{bmatrix} -0.42 & -0.07 & -0.10 \end{bmatrix},$$

$$C_{F2} = \begin{bmatrix} -0.31 & 0.05 & -0.27 \end{bmatrix},$$

$$D_{F1} = 1.14, \; D_{F2} = 0.04.$$

The effectiveness of the designed filter can be verified by observing the filtering error curve shown in Fig. 7.4.

7.6 CONCLUSION

The robust H_∞ filtering problem for discrete-time uncertain switched systems is investigated via the N-step ahead LF approach. Such an approach can achieve better filtering performance, whereas as a cost, the restriction on the switching law is required to be more critical. To make a trade-off between filtering performance and switching law, the MDADT switching is introduced to reduce the ADT bound such that a more general result, i.e., better filtering capability with smaller ADT bound, can be achieved.

References

[1] Y. Wei, J. Qiu, P. Shi, et al., Fixed-order piecewise-affine output feedback controller for fuzzy-affine-model-based nonlinear systems with time-varying delay, IEEE Transactions on Circuits and Systems I: Regular Papers 64 (4) (2017) 945–958.

[2] Y. Wei, J. Qiu, H.R. Karimi, Fuzzy-affine-model-based memory filter design of nonlinear systems with time-varying delay, IEEE Transactions on Fuzzy Systems 26 (2) (2017) 504–517.

[3] A. Hocine, M. Chadli, H.R. Karimi, A structured filter for Markovian switching systems, International Journal of Systems Science 45 (7) (2014) 1518–1527.

[4] H. Li, Y. Gao, L. Wu, et al., Fault detection for TS fuzzy time-delay systems: delta operator and input–output methods, IEEE Transactions on Cybernetics 45 (2) (2015) 229–241.

[5] Y. Gao, F. Xiao, J. Liu, et al., Distributed soft fault detection for interval type-2 fuzzy-model-based stochastic systems with wireless sensor networks, IEEE Transactions on Industrial Informatics 15 (1) (2019) 334–347.

[6] Z. Gao, Kalman filters for continuous-time fractional-order systems involving fractional-order colored noises using Tustin generating function, International Journal of Control, Automation, and Systems 16 (3) (2018) 1049–1059.

[7] S. Beidaghi, A.A. Jalali, A.K. Sedigh, New H_2 filtering for descriptor systems: singular and normal filters, International Journal of Control, Automation, and Systems 15 (1) (2017) 160–168.

[8] X. Li, J. Lam, H. Gao, et al., H_∞ and H_2 filtering for linear systems with uncertain Markov transitions, Automatica 67 (2016) 252–266.

[9] D. Du, B. Jiang, P. Shi, et al., H_∞ filtering of discrete-time switched systems with state delays via switched Lyapunov function approach, IEEE Transactions on Automatic Control 52 (8) (2007) 1520–1525.

[10] H. Liu, P. Shi, H.R. Karimi, et al., Finite-time stability and stabilisation for a class of nonlinear systems with time-varying delay, International Journal of Systems Science 47 (6) (2016) 1433–1444.

[11] E. Tian, W.K. Wong, D. Yue, et al., H_∞ filtering for discrete-time switched systems with known sojourn probabilities, IEEE Transactions on Automatic Control 60 (9) (2015) 2446–2451.

[12] K. Hu, J. Yuan, Improved robust H_∞ filtering for uncertain discrete-time switched systems, IET Control Theory & Applications 3 (3) (2009) 315–324.

[13] M. Wag, J. Qiu, M. Chadli, et al., A switched system approach to exponential stabilization of sampled-data T–S fuzzy systems with packet dropouts, IEEE Transactions on Cybernetics 46 (12) (2016) 3145–3156.

[14] P. Zhao, R. Nagamune, Switching LPV controller design under uncertain scheduling parameters, Automatica 76 (2017) 243–250.

[15] L. Zhang, E.K. Boukas, P. Shi, Exponential H_∞ filtering for uncertain discrete-time switched linear systems with average dwell time: a μ-dependent approach, International Journal of Robust and Nonlinear Control 18 (11) (2008) 1188–1207.

[16] D. Wang, P. Shi, J. Wang, et al., Delay-dependent exponential H_∞ filtering for discrete-time switched delay systems, International Journal of Robust and Nonlinear Control 22 (13) (2012) 1522–1536.

[17] L. Zhang, N. Cui, M. Liu, et al., Asynchronous filtering of discrete-time switched linear systems with average dwell time, IEEE Transactions on Circuits and Systems I: Regular Papers 58 (5) (2011) 1109–1118.

[18] J. Lian, C. Mu, P. Shi, Asynchronous H_∞ filtering for switched stochastic systems with time-varying delay, Information Sciences 224 (2013) 200–212.

[19] X. Zhao, L. Zhang, P. Shi, et al., Stability and stabilization of switched linear systems with mode-dependent average dwell time, IEEE Transactions on Automatic Control 57 (7) (2012) 1809–1815.

[20] H. Zhang, D. Xie, H. Zhang, et al., Stability analysis for discrete-time switched systems with unstable subsystems by a mode-dependent average dwell time approach, ISA Transactions 53 (4) (2014) 1081–1086.

[21] J. Cheng, H. Zhu, S. Zhong, et al., Finite-time filtering for switched linear systems with a mode-dependent average dwell time, Nonlinear Analysis: Hybrid Systems 15 (2015) 145–156.

[22] X. Zhao, H. Liu, Z. Wang, Weighted H_∞ performance analysis of switched linear systems with mode-dependent average dwell time, International Journal of Systems Science 44 (11) (2013) 2130–2139.

[23] H. Liu, X. Zhao, Finite-time H_∞ control of switched systems with mode-dependent average dwell time, Journal of the Franklin Institute 351 (3) (2014) 1301–1315.

[24] J.W. Wen, S.K. Nguang, P. Shi, X.D. Zhao, Stability and H_∞ control of discrete-time switched systems via one-step ahead Lyapunov function approach, IET Control Theory & Applications 12 (8) (2018) 1141–1147.

[25] A. Nasiri, S.K. Nguang, A. Swain, et al., Robust output feedback controller design of discrete-time Takagi–Sugeno fuzzy systems: a non-monotonic Lyapunov approach, IET Control Theory & Applications 10 (5) (2016) 545–553.

[26] S.F. Derakhshan, A. Fatehi, Non-monotonic Lyapunov functions for stability analysis and stabilization of discrete time Takagi–Sugeno fuzzy systems, International Journal of Innovative Computing, Information & Control 10 (4) (2014) 1567–1586.

[27] A. Kruszewski, R. Wang, T.M. Guerra, Nonquadratic stabilization conditions for a class of uncertain nonlinear discrete time T-S fuzzy models: a new approach, IEEE Transactions on Automatic Control 53 (2) (2008) 606–611.

[28] J. Wen, S.K. Nguang, P. Shi, et al., Robust H_∞ control of discrete-time nonhomogenous Markovian jump systems via multistep Lyapunov function approach, IEEE Transactions on Systems, Man, and Cybernetics: Systems 47 (7) (2017) 1439–1450.

[29] Y. Xie, J.W. Wen, S.K. Nguang, L. Peng, Stability, l_2-gain and robust H_∞ control for switched systems via N-step ahead Lyapunov function approach, IEEE Access 5 (1) (2017) 26400–26408.

[30] M.S. Ali, K. Meenakshi, R. Vadivel, et al., Robust H_∞ performance of discrete-time neural networks with uncertainty and time-varying delay, International Journal of Control, Automation, and Systems 16 (4) (2018) 1637–1647.

8

Dissipative Dynamic Output Feedback Control for Switched Systems via Multistep Lyapunov Function Approach

8.1 INTRODUCTION

The basic problem in the control theory is exploring control laws such that desired control performance can be guaranteed [1]. Among various control strategies, state feedback and output feedback are most well known, where the former has a critical requirement on measured states [2], and the latter is relatively easy in application due to the accessibility of outputs [3]. Therefore, burgeoning results on output feedback have been reported in the literature, and such a control approach can be roughly divided into static and dynamic ones. The structure of static output feedback (SOF) is simple, and its implementation is convenient [4,5]. However, it is usually difficult to formulate the stabilization problem of SOF in terms of standard linear matrix inequalities (LMIs), especially for complex systems. Even though the bilinear matrix inequality (BMI) formulation [6] is developed to deal with such a problem, but as mentioned in [7], BMI is nonconvex and NP-hard. Unlike SOF, which still remains a difficult solvability problem [8], dynamic output feedback (DOF) control is relatively mature in theoretical design.

In terms of engineering applications, DOF drives extensive investigations. For instance, in networked control systems (NCSs), the plant and controller are spatially separated, where the control-loop is physically linked by a communication network, and it may lead to network-induced delays and data packet dropouts [9]. Generally, obtaining an accurate

Non-Monotonic Approach to Robust H∞ Control of Multi-Model Systems
https://doi.org/10.1016/B978-0-12-814868-6.00014-9

knowledge of the plant state to achieve a routine state feedback control is quite challenging since some inner state variables cannot be directly measured by sensor nodes. Then a natural consideration is designing observer-based controllers. However, the estimated error is difficult to be obtained due to the time delay between the control input at the plant side and observer side. Therefore, DOF can be applied to NCS with transmission time delay and data packet dropouts [10]. Another typical example of the DOF application is wireless control system (WCS) with control network. In a WCS, it is very complicated to share local control of each node when constructing an observer. That is to say, observer-based controllers need a unified control input for the purpose of observing error computing. Therefore, it is reasonable to design DOF controllers when each node can design and perform local control without a unified control input knowledge.

On the other hand, dissipativity theory provides a framework for the analysis and synthesis of control systems [11]. A general theory of dissipative dynamical systems was first outlined in [12], where the dissipativity was characterized by storage functions and supply rates. A state-space approach for linear dissipative dynamical systems was further addressed in [13]. Extension results to the case of affine nonlinear systems were carried out in [14,15]. Some important advances have been reported for the dissipativity property of switched systems. In [16], novel cross-supply rates among subsystems were introduced to describe the changes of storage function with inactive subsystems. The concept of decomposable dissipativity was firstly proposed in [17], where the nonpositive condition of the supply rate was relaxed by the decomposition of supply rate. By imposing decreasing condition on storage functions at switching instants, the dissipativity and induced Lyapunov stability of switched systems was established [18]. In general cases, the dissipativity of switched systems is established by constructing a Lyapunov function (LF). Such a decreasing requirement on LFs at switching instants was replaced by an increasing one in [19]. However, the LFs in existing literature are required to be monotonically decreasing within subsystems, which motivates us to explore a nonmonotonic LF approach to further reduce such a conservatism.

For the average dwell-time (ADT) switched systems [20,21], the LFs are allowed to increase at the switching instant rather than within the subsystem. Some pioneering works on the nonmonotonic LF have been reported. For instance, nonmonotonic LFs are firstly developed in [22] to stabilize nonlinear discrete-time switched systems such that less conservatism results can be obtained. Based on this fundamental work, a one-step ahead LF that can ensure globally uniformly asymptotically stable (GUAS) property for arbitrarily switched systems with further less conservativeness was formulated in [23]. A similar method was adopted to reduce conservatism in a class of nonhomogenous Markovian jump systems by allowing LF to increase during the period of several sampling time steps ahead of the current time within each jump mode [24]. However, to our best knowledge, the exploration of nonmonotonic results for ADT switching systems still remains open, especially when the N-step ahead scenario is taken into account [25]. The essential difficulty is constructing an exponential damping law of the LF decreasing points, i.e., to find the joint point between the switching interval and the predictive horizon.

Therefore, in this chapter, from the conservatism reduction point of view, a nonmonotonic dissipative DOF control approach is developed for a class of switched systems. The underlying LF is allowed to increase both at the switching instants and during the running time of each subsystem. The main contributions and novelty of this chapter are summarized as follows:

1) A multistep Lyapunov function (MLF) approach is developed for conservatism reduction of dissipative analysis such that the dissipative region can be enlarged with guaranteed stability and, specifically, smaller H_∞ level.

2) A less conservative DOF controller design is further developed based on the analysis results, where the LF matrix is formulated without structural constraints to avoid equality constraint.

Notations: $\mathbb{R}^n$ represents the n-dimensional Euclidean space, and $\mathbb{Z}^+$ is the set of nonnegative integers. The superscript "T" stands for matrix transposition, the symbol $*$ is an ellipsis for symmetric terms, and $diag\{\cdots\}$ stands for a block-diagonal matrix; $l_2[0, \infty]$ is the space of square-summable infinite sequences for exogenous noise $w_k \in l_2[0, \infty]$. The notation $P > 0$ ($P \geq 0$) means that P is a positive definite (semidefinite) matrix; $\aleph_{1,r}$ represents a set takeing values in an integer set $\{1, 2, 3, \ldots, r\}$, where $r \in \mathbb{Z}^+$. Let R be a negative definite matrix and define $-R \triangleq R_-$. Then there exists a matrix $\left(R_-^{1/2}\right)^2 \geq 0$ such that $-R = \left(R_-^{1/2}\right)^2$. A function $\beta : [0, \infty) \times [0, \infty) \rightarrow [0, \infty)$ is said to be of class $\mathcal{KL}$ if $\beta(\cdot, t)$ is of class $\mathcal{K}$ for each fixed $t \geq 0$ and $\beta(s, t)$ decreases to 0 as $t \rightarrow \infty$ for each fixed $s \geq 0$.

8.2 PRELIMINARIES AND PROBLEM FORMULATION

Consider the following discrete-time switched system:

$$\begin{cases} x_{k+1} = (A_{\sigma(k)} + \Delta A_{\sigma(k)}^\eta)x_k + B_{\sigma(k)}u_k + B_{\sigma(k)\omega}\omega_k, \\ y_k = L_{\sigma(k)}x_k + D_{\sigma(k)\omega}\omega_k, \\ z_k = C_{\sigma(k)}x_k + D_{\sigma(k)}u_k, \end{cases} \tag{8.1}$$

where $x(k) \in \mathbb{R}^{n_x}$ is the state, $y(k) \in \mathbb{R}^{n_y}$ and $z(k) \in \mathbb{R}^{n_z}$ are the controlled output and measured output, respectively, $\omega(k) \in \mathbb{R}^{n_\omega}$ represents the disturbance belonging to $l_2[0, \infty]$. The switching signal $\sigma(k)$ takes its value in the finite set $\aleph_{1,r}$, and $\sigma(k) = i$ means that the ith subsystem is activated. The matrices A_i, B_i, $B_{i\omega}$, C_i, D_i, L_i, and $D_{i\omega}$ are constant matrices of appropriate dimensions. Furthermore, ΔA_i^η represents norm-bounded uncertainties decomposed as $E_i F_\eta H_i$, where E_i and H_i are known matrices, which characterize the structure of the uncertainties; F_η ($\eta \in [k, k+N)$) models the time-varying element satisfying $F_\eta^T F_\eta < I$.

Definition 8.1. [26] For a finite switching sequence $\sigma(k)$ and any $K > k > k_0$, $N_\sigma(K, k)$ stands for the switching number of $\sigma(k)$ over the interval $[k, K)$. If there exists $N_0 > 0$ (the chatter bounds) such that $N_\sigma(K, k) \leq N_0 + (K - k)/\tau_a$, then we call τ_a the ADT of the system.

In this chapter, the DOF controller is formulated in the form

$$\begin{bmatrix} \hat{x}_{k+1} \\ u_k \end{bmatrix} = \begin{bmatrix} \hat{A}_i & \hat{B}_i \\ \hat{C}_i & 0 \end{bmatrix} \begin{bmatrix} \hat{x}_k \\ y_k \end{bmatrix}, \tag{8.2}$$

where $\hat{x}_k$ is the state vector of the controller, and $\hat{A}_i$, $\hat{B}_i$, and $\hat{D}_i$ are controller matrices of appropriate dimensions to be designed. By constructing an augmented state vector $\tilde{x}(k) = [\; x^T(k) \quad \hat{x}^T(k)\;]^T$ it follows from (8.1) and (8.2) that

$$\begin{cases} \tilde{x}_{k+1} = \tilde{A}_i \tilde{x}_k + \tilde{B}_i \omega_k, \\ z_k = \tilde{C}_i \tilde{x}_k, \end{cases} \tag{8.3}$$

where

$$\tilde{A}_i = \begin{bmatrix} A_i + \Delta A_i^\eta & B_i \hat{C}_i \\ \hat{B}_i L_i & \hat{A}_i \end{bmatrix}, \tilde{B}_i = \begin{bmatrix} B_{i\omega} \\ \tilde{B}_i D_{i\omega} \end{bmatrix}, \tilde{C}_i = \begin{bmatrix} C_i & D_i \hat{C}_i \end{bmatrix}.$$

Now, introduce the quadratic energy supply function $E(\cdot, \cdot)$ associated with the above augmented system:

$$E(\omega, z) \triangleq z^T R z + z^T S \omega + \omega^T Q \omega, \tag{8.4}$$

where $Q = Q^T > 0$, $S = S^T$, and $R \leq 0$ are real matrices of appropriate dimensions.

Definition 8.2. [11] Under the zero initial condition $x(0) = 0$, system (8.3) with energy supply (8.4) is strictly (Q, S, R)-dissipative if there exists a sufficiently small scalar β such that the following inequality holds for all $T \geq 0$ and $\omega_k \in l_2[0, T]$:

$$\sum_{k=0}^{T} E(\omega_k, z_k) \geq \beta \sum_{k=0}^{T} w_k^T w_k. \tag{8.5}$$

In the following, the main objective is to derive general sufficient conditions that guarantee the robust GUAS and strict dissipativity of system (8.3).

8.3 DISSIPATIVE ANALYSIS

In this section, we address an MLF technique and formulate dissipativity conditions by a set of LMIs.

Theorem 8.1. *Let $0 < \alpha < 1$ and $\mu > 1$ be given constants, and let Q, S, R be given matrices with Q and R symmetric. The switched system (8.3) is GUAS and strictly (Q, S, R)-dissipative when $\omega_k = 0$ if the ADT of switching signal $\sigma(k)$ satisfies $\tau_a > \tau_a^* = -N \ln \mu / \ln(1 - \alpha)$ and there exists a set of symmetric matrices $Z_{i,f}$ $(f \in \aleph_{1,N-1})$ and $P_i > 0$ such that the following LMIs hold:*

for $N = 2$,

$$
\begin{bmatrix}
-P_i & P_i \tilde{A}_i & P_i \tilde{B}_i & 0 \\
* & -Z_{i,1} & -\tilde{C}_i^T S & \tilde{C}_i^T R_-^{1/2} \\
* & * & -Q & 0 \\
* & * & * & -I
\end{bmatrix} \le 0,
\tag{8.6}
$$

$$
\begin{bmatrix}
-Z_{i,1} & Z_{i,1} \tilde{A}_i & Z_{i,1} \tilde{B}_i & 0 \\
* & (\alpha - 1) P_i & -\tilde{C}_i^T S & \tilde{C}_i^T R_-^{1/2} \\
* & * & -Q & 0 \\
* & * & * & -I
\end{bmatrix} \le 0,
\tag{8.7}
$$

for $N \ge 3$ and $f \in \aleph_{2,N-1}$,

$$
\begin{bmatrix}
-P_i & P_i \tilde{A}_i & P_i \tilde{B}_i & 0 \\
* & -Z_{i,N-1} & -\tilde{C}_i^T S & \tilde{C}_i^T R_-^{1/2} \\
* & * & -Q & 0 \\
* & * & * & -I
\end{bmatrix} \le 0,
\tag{8.8}
$$

$$
\begin{bmatrix}
-Z_{i,N-f+1} & Z_{i,N-f+1} \tilde{A}_i & Z_{i,N-f+1} \tilde{B}_i & 0 \\
* & -Z_{i,N-f} & -\tilde{C}_i^T S & \tilde{C}_i^T R_-^{1/2} \\
* & * & -Q & 0 \\
* & * & * & -I
\end{bmatrix} \le 0,
\tag{8.9}
$$

$$\begin{bmatrix} -Z_{i,1} & Z_{i,1}\tilde{A}_i & Z_{i,1}\tilde{B}_i & 0 \\ * & (\alpha-1)P_i & -\tilde{C}_i^T S & \tilde{C}_i^T R_-^{1/2} \\ * & * & -Q & 0 \\ * & * & * & -I \end{bmatrix} \leq 0, \tag{8.10}$$

and

$$P_i \leq \mu P_j \tag{8.11}$$

for all $\sigma(k_l) = i$ and $\sigma(k_l^-) = j$.

Proof. Since the derivation for the cases of $N = 2$ and $N \geq 3$ are in a similar routine, we give the proof for $N \geq 3$ in detail and only some brief notes for $N = 2$.

Before dissipative analysis, we aim at presenting sufficient conditions for the GUAS of system (8.3), which can be implied by Theorem 8.1.

The LF in this chapter is taken as $V_i(\tilde{x}_k) = \tilde{x}_k^T P_i \tilde{x}_k$, and its N-sample variation is

$$\Delta V_{i,N}(\tilde{x}_k) = V_i(\tilde{x}_{k+N}) - V_i(\tilde{x}_k). \tag{8.12}$$

Let us further consider the auxiliary quadratic function

$$V_i'(\tilde{x}_{k+f}) = \tilde{x}_{k+f}^T Z_{i,f} \tilde{x}_{k+f}, \ f \in \aleph_{1,N-1}.$$

In view of conditions (8.8)–(8.10), we obtain the following LMIs by performing some basic matrix manipulations:

$$\begin{bmatrix} -P_i & P_i\tilde{A}_i \\ * & -Z_{i,N-1} \end{bmatrix} \leq 0, \tag{8.13}$$

$$\begin{bmatrix} -Z_{i,N-f+1} & -Z_{i,N-f+1}\tilde{A}_i \\ * & -Z_{i,N-f} \end{bmatrix} \leq 0, f \in \aleph_{2,N-1}, \tag{8.14}$$

$$\begin{bmatrix} -Z_{i,1} & Z_{i,1}\tilde{A}_i \\ * & (\alpha_i-1)P_i \end{bmatrix} \leq 0. \tag{8.15}$$

The inequalities

$$V_i(\tilde{x}_{k+N}) - V_i'(\tilde{x}_{k+N-1}) \leq 0, \tag{8.16}$$

$$V_i'(\tilde{x}_{k+N-f+1}) - V_i'(\tilde{x}_{k+N-f}) \leq 0, f \in \aleph_{2,N-1}, \tag{8.17}$$

$$V_i'(\tilde{x}_{k+1}) + (\alpha_i-1)V_i(\tilde{x}_k) \leq 0 \tag{8.18}$$

can be inferred from (8.13)–(8.15) when we assume the zero disturbance input to the system.

It follows from LMIs (8.16)–(8.18) that

$$V_i(\tilde{x}_{k+N}) - V_i(\tilde{x}_k) \leq -\alpha_i V_i(\tilde{x}_k). \tag{8.19}$$

Assuming that the switching behavior is much slower than the sampling time step, it follows from (8.19) that

$$V_{\sigma(k)}(x_k) \leq (1 - \alpha)^{\frac{k-k_l}{N}} V_{\sigma(k)}(x_{k_l}). \tag{8.20}$$

Denote by k_l the switching times, and when condition (8.11) holds, it results in

$$V_i(x_{k_l}) \leq \mu V_j(x_{k_l}). \tag{8.21}$$

Combing (8.20) and (8.21), we obtain

$$V_{\sigma(k)}(x_k) \leq \mu^{N_0} [\mu^{\frac{1}{\tau_a}} (1 - \alpha)^{\frac{1}{N}}]^{k-k_0} V_{\sigma(k_0)}(x_{k_0}). \tag{8.22}$$

If the ADT satisfies $\tau_a > \tau_a^* = -N \ln \mu / \ln(1 - \alpha)$, then we can directly conclude that $V_{\sigma(k)}(x_k) \leq \mu^{N_0} V_{\sigma(k_0)}(x_{k_0})$. Therefore, conditions (8.8)–(8.10) guarantee the GUAS of system (8.3) without disturbance for $N \geq 3$.

Then, assuming that system (8.3) is under the zero-initial condition, we are in the position to perform dissipative analysis.

Due to the nonmonotonic feature of MLF, we introduce a set of auxiliary matrices $Z_{i,f}$ to construct connection from k to $k + N$. Along the trajectory of system (8.3), the inequalities

$$V_i(\tilde{x}_{k+N}) - V_i'(\tilde{x}_{k+N-1}) - E(\omega_{k+N-1}, z_{k+N-1}) \leq 0, \tag{8.23}$$

$$V_i'(\tilde{x}_{k+N-f+1}) - V_i'(\tilde{x}_{k+N-2}) - E(\omega_{k+N-2}, z_{k+N-2}) \leq 0, \, f \in \aleph_{2.N-1}, \tag{8.24}$$

$$V_i'(\tilde{x}_{k+1}) + (\alpha - 1)V_i(\tilde{x}_k) - E(\omega_k, z_k) \leq 0 \tag{8.25}$$

can be directly inferred from (8.8)–(8.10).

It follows from conditions (8.23)–(8.25) that

$$\Delta V_{i,N}(\tilde{x}_k) + \alpha V_i(\tilde{x}_k) - \sum_{f=0}^{N-1} E(\omega_{k+f}, z_{k+f}) \leq 0. \tag{8.26}$$

Under the zero initial conditions $x_0 = x_{-1} = x_{-2} = \cdots = x_{-N+1} = 0$, we have

$$\Delta V_{i,N}(\tilde{x}_k)\} = V_i(\tilde{x}_{T+1}) + V_i(\tilde{x}_{T+2}) + \cdots + V_i(\tilde{x}_{T+N}).$$

As $T \to \infty$, it straightforwardly results in

$$NV_i(\tilde{x}_\infty) - N\sum_{k=0}^{T} E(\omega_k, z_k) \le \sum_{k=0}^{\infty} \{-\alpha_i V_i(\tilde{x}_k)\}.$$

It is obvious that $\sum_{k=0}^{T} E(\omega_k, z_k) \ge 0$ and there exists a sufficiently small β such that $\sum_{k=0}^{T} E(\omega_k, z_k) \ge \beta \sum_{k=0}^{T} \omega_k^T \omega_k$. The proof is completed. □

Remark 8.1. It should be noted that $N = 1$ is a particular case of MLF since it is required to decrease monotonically within subsystems. The dissipative analysis for $N = 1$ is omitted since the derivation is quite easy. Furthermore, note that N has direct impact on the number of the inequality constraints. For $N = 2$, conditions (8.16) and (8.18) can be obtained from (8.6) and (8.7), respectively. The corresponding dissipative analysis can be performed by following the same lines of the proof for $N = 3$.

Remark 8.2. (Q, S, R)-dissipativity covers various forms of dissipativity, such as H_∞ performance and passivity, as particular cases.

1. When $R = -1$, $S = 0$, and $Q = \gamma^2$, strict (Q, S, R)-dissipativity (8.5) reduces to an H_∞ performance requirement.
2. When $R = 0$, $S = 1/2$, and $Q = -\eta$ ($\eta \ge 0$), condition (8.4) implies the robust passivity.

8.4 ROBUST DISSIPATIVE DOF CONTROL

In this section, robust dissipative DOF controller design is developed based on the dissipative analysis.

Theorem 8.2. *Let $0 < \alpha < 1$, $\mu > 1$, and $\varepsilon_i > 1$ be given constants, and let Q, S, R be given matrices with Q and R symmetric. For ADT of the switching signal $\sigma(k)$ satisfies $\tau_a > \tau_a^* = -N \ln \mu / \ln(1 - \alpha)$ if there exist matrices T_i, V_i, A_{icc} and a set of symmetric matrices $W_i^{11} > 0$, $W_i^{12} > 0$, $W_i^{22} > 0$, $X_i, Y_i, K_{i,f}^{11}, K_{i,f}^{12}, K_{i,f}^{22}$ ($f \in \aleph_{1,N-1}$) such that the following LMIs hold:*

for $N = 2$,

$$\begin{bmatrix} \Psi^{11} & \Psi^{12} & \Psi^{13} & 0 & \Psi^{15} & 0 \\ * & \Xi^{22} & \Psi^{23} & \Psi^{24} & 0 & \Psi^{26} \\ * & * & -Q & 0 & 0 & 0 \\ * & * & * & -I & 0 & 0 \\ * & * & * & * & -\varepsilon_i^{-1}I & 0 \\ * & * & * & * & * & -\varepsilon_i I \end{bmatrix} \le 0, \qquad (8.27)$$

$$
\begin{bmatrix}
\Xi^{11} & \Psi^{12} & \Psi^{13} & 0 & \Psi^{15} & 0 \\
* & \Psi^{22} & \Psi^{23} & \Psi^{24} & 0 & \Psi^{26} \\
* & * & -Q & 0 & 0 & 0 \\
* & * & * & -I & 0 & 0 \\
* & * & * & * & -\varepsilon_i^{-1}I & 0 \\
* & * & * & * & * & -\varepsilon_i I
\end{bmatrix} \leq 0, \tag{8.28}
$$

for $N = 3$,

$$
\begin{bmatrix}
\Psi^{11} & \Psi^{12} & \Psi^{13} & 0 & \Psi^{15} & 0 \\
* & \Theta^{22} & \Psi^{23} & \Psi^{24} & 0 & \Psi^{26} \\
* & * & -Q & 0 & 0 & 0 \\
* & * & * & -I & 0 & 0 \\
* & * & * & * & -\varepsilon_i^{-1}I & 0 \\
* & * & * & * & * & -\varepsilon_i I
\end{bmatrix} \leq 0, \tag{8.29}
$$

$$
\begin{bmatrix}
\Upsilon^{11} & \Psi^{12} & \Psi^{13} & 0 & \Psi^{15} & 0 \\
* & \Upsilon^{22} & \Psi^{23} & \Psi^{24} & 0 & \Psi^{26} \\
* & * & -Q & 0 & 0 & 0 \\
* & * & * & -I & 0 & 0 \\
* & * & * & * & -\varepsilon_i^{-1}I & 0 \\
* & * & * & * & * & -\varepsilon_i I
\end{bmatrix} \leq 0, \tag{8.30}
$$

$$
\begin{bmatrix}
\Xi^{11} & \Psi^{12} & \Psi^{13} & 0 & \Psi^{15} & 0 \\
* & \Psi^{22} & \Psi^{23} & \Psi^{24} & 0 & \Psi^{26} \\
* & * & -Q & 0 & 0 & 0 \\
* & * & * & -I & 0 & 0 \\
* & * & * & * & -\varepsilon_i^{-1}I & 0 \\
* & * & * & * & * & -\varepsilon_i I
\end{bmatrix} \leq 0, \tag{8.31}
$$

and

$$
P_i \leq \mu P_j
$$

for $\sigma(k_l) = i$ and $\sigma(k_l^-) = j$, where

$$
\Psi^{11} = \begin{bmatrix}
-2Y_i + W_i^{11} & -2I + W_i^{12} \\
* & -2X_i + W_i^{22}
\end{bmatrix},
$$

$$
\Psi^{12} = \begin{bmatrix}
A_i Y_i + B_i T_i & A_i \\
A_{icc} & X_i A_i + V_i L_i
\end{bmatrix},
$$

$$\Upsilon^{11} = \begin{bmatrix} -2Y_i + \mathrm{K}^{11}_{i,N-f+1} & -2I + \mathrm{K}^{12}_{i,N-f+1} \\ * & -2X_i + \mathrm{K}^{22}_{i,N-f+1} \end{bmatrix},$$

$$\Upsilon^{22} = \begin{bmatrix} -\mathrm{K}^{11}_{i,N-f} & -\mathrm{K}^{12}_{i,N-f} \\ * & -\mathrm{K}^{22}_{i,N-f} \end{bmatrix},$$

$$\Psi^{24} = \begin{bmatrix} Y_i C_i^T R_-^{1/2} + T_i^T D_i^T R_-^{1/2} \\ C_i^T R_-^{1/2} \end{bmatrix}, \quad \Xi^{11} = \begin{bmatrix} -2Y_i + \mathrm{K}^{11}_{i,1} & -2I + \mathrm{K}^{12}_{i,1} \\ * & -2X_i + \mathrm{K}^{22}_{i,1} \end{bmatrix},$$

$$\Psi^{13} = \begin{bmatrix} B_{i\omega} \\ X_i B_{i\omega} + V_i D_{i\omega} \end{bmatrix}, \quad \Psi^{15} = \begin{bmatrix} E_i & 0 \\ X_i E_i & 0 \end{bmatrix}, \quad \Psi^{26} = \begin{bmatrix} Y_i H_i & 0 \\ H_i & 0 \end{bmatrix},$$

$$\Psi^{22} = \begin{bmatrix} (\alpha-1)W_i^{11} & (\alpha-1)W_i^{12} \\ * & (\alpha-1)W_i^{22} \end{bmatrix}, \quad \Psi^{23} = \begin{bmatrix} -Y_i C_i^T S - T_i^T D_i^T S \\ -C_i^T S \end{bmatrix},$$

$$\Xi^{22} = \begin{bmatrix} (\alpha-1)\mathrm{K}^{11}_{i,1} & (\alpha-1)\mathrm{K}^{12}_{i,1} \\ * & (\alpha-1)\mathrm{K}^{22}_{i,1} \end{bmatrix}, \quad \Theta^{22} = \begin{bmatrix} -\mathrm{K}^{11}_{i,N-1} & -\mathrm{K}^{12}_{i,N-1} \\ * & -\mathrm{K}^{22}_{i,N-1} \end{bmatrix}.$$

Then system (8.3) is GUAS with the controller of the form (8.2), and the control matrices can be obtained as

$$\hat{A}_i = (Y_i^{-1} - X_i)^{-1}(A_{icc} - X_i A_i Y_i - V_i L_i Y_i - X_i B_i T_i)Y_i^{-1},$$
$$\hat{B}_i = (Y_i^{-1} - X_i)^{-1}V_i, \hat{C}_i = T_i Y_i^{-1}.$$

Proof. In this section, we still focus on the derivation for $N = 3$.

For matrix transformation, we introduce the matrices

$$G_i = \begin{bmatrix} X_i & Y_i^{-1} - X_i \\ * & X_i - Y_i^{-1} \end{bmatrix},$$

where both X_i and Y_i are symmetric. Pre- and postmultiply $diag\{G_i P_i^{-1}, I, I, I\}$ and $diag\{P_i^{-1}G_i^T, I, I, I\}$ to (8.9), respectively. Since $-G_i^T - G_i + P_i \le -G_i P_i^{-1}G_i^T$, we obtain

$$\Gamma_1 = \begin{bmatrix} -G_i^T - G_i + P_i & G_i \tilde{A}_i & G_i \tilde{B}_i & 0 \\ * & -Z_{i,N-1} & -\tilde{C}_i^T S & \tilde{C}_i^T R_-^{1/2} \\ * & * & -Q & 0 \\ * & * & * & -I \end{bmatrix} \le 0. \qquad (8.32)$$

Taking the uncertainties into account, we denote

$$\tilde{A}_i = \tilde{A}_i^1 + \Delta\tilde{A}_i = \begin{bmatrix} A_i & B_i\hat{C}_i \\ \hat{B}_i L_i & \hat{A}_i \end{bmatrix} + \begin{bmatrix} \bar{E}_i\bar{F}_\eta\bar{H}_i & 0 \\ 0 & 0 \end{bmatrix},$$

$$\bar{E}_i = \begin{bmatrix} E_i & 0 \\ 0 & 0 \end{bmatrix}, \bar{F}_\eta = \begin{bmatrix} F_\eta & 0 \\ 0 & 0 \end{bmatrix}, \bar{H}_i = \begin{bmatrix} H_i & 0 \\ 0 & 0 \end{bmatrix},$$

$$\tilde{E}_i = \begin{bmatrix} 0 & G_i\bar{E}_i & 0 & 0 \\ 0 & 0 & 0 & 0 \\ 0 & 0 & 0 & 0 \\ 0 & 0 & 0 & 0 \end{bmatrix}, \tilde{F}_\eta = \begin{bmatrix} 0 & 0 & 0 & 0 \\ 0 & \bar{F}_\eta & 0 & 0 \\ 0 & 0 & 0 & 0 \\ 0 & 0 & 0 & 0 \end{bmatrix}, \tilde{H}_i = \begin{bmatrix} 0 & 0 & 0 & 0 \\ 0 & \bar{H}_i & 0 & 0 \\ 0 & 0 & 0 & 0 \\ 0 & 0 & 0 & 0 \end{bmatrix}.$$

Subsequently, (8.32) can be decomposed as

$$\Gamma_1 = \Gamma_0 + \tilde{E}_i\tilde{F}_\eta\tilde{H}_i + \left(\tilde{E}_i\tilde{F}_\eta\tilde{H}_i\right)^T,$$

where

$$\Gamma_0 = \begin{bmatrix} -G_i^T - G_i + P_i & G_i\tilde{A}_i^1 & G_i\tilde{B}_i & 0 \\ * & -Z_{i,N-1} & -\tilde{C}_i^T S & \tilde{C}_i^T R_-^{1/2} \\ * & * & -Q & 0 \\ * & * & * & -I \end{bmatrix} \le 0.$$

Since $\Gamma_0 + \tilde{E}_i\tilde{F}_\eta\tilde{H}_i + \left(\tilde{E}_i\tilde{F}_\eta\tilde{H}_i\right)^T \le \varepsilon_i\tilde{E}_i\tilde{E}_i^T + \varepsilon_i^{-1}\tilde{H}_i^T\tilde{H}_i$, we can obtain the LMI

$$\begin{bmatrix} -G_i^T - G_i + P_i & G_i\tilde{A}_i & G_i\tilde{B}_i & 0 & G_i\bar{E}_i & 0 \\ * & -Z_{i,N-1} & -\tilde{C}_i^T S & \tilde{C}_i^T R_-^{1/2} & 0 & \bar{H}_i \\ * & * & -Q & 0 & 0 & 0 \\ * & * & * & -I & 0 & 0 \\ * & * & * & * & -\varepsilon_i^{-1}I & 0 \\ * & * & * & * & * & -\varepsilon_i I \end{bmatrix} \le 0 \tag{8.33}$$

by employing Schur's complement.

Performa congruence transformation to (8.33) by $diag\{J_i^T, J_i^T, I, I\}$ on the left and $diag\{J_i, J_i, I, I\}$ on the right with $J_i = \begin{bmatrix} Y_i & I \\ Y_i & 0 \end{bmatrix}$. Then, condition (8.33) can be transformed into

$$
\begin{bmatrix}
J_i^T(-G_i^T - G_i + P_i)J_i & J_i^T G_i \tilde{A}_i J_i & J_i^T G_i \tilde{B}_i & 0 & J_i^T G_i \bar{E}_i & 0 \\
* & -J_i^T Z_{i,N-1} J_i & -J_i^T \tilde{C}_i^T S & J_i^T \tilde{C}_i^T R_-^{1/2} & 0 & J_i^T \bar{H}_i \\
* & * & -Q & 0 & 0 & 0 \\
* & * & * & -I & 0 & 0 \\
* & * & * & * & -\varepsilon_i^{-1} I & 0 \\
* & * & * & * & * & -\varepsilon_i I
\end{bmatrix} \le 0.
$$

(8.34)

Setting

$$
J_i^T P_i J_i \triangleq W_i = \begin{bmatrix} W_i^{11} & W_i^{12} \\ * & W_i^{22} \end{bmatrix} \text{ and } J_i^T Z_{i,f} J_i \triangleq K_i = \begin{bmatrix} K_{i,f}^{11} & K_{i,f}^{12} \\ * & K_{i,f}^{22} \end{bmatrix},
$$

we can readily obtain

$$
J_i^T(-G_i^T - G_i + P_i)J_i = \begin{bmatrix} -2Y_i + W_i^{11} & -2I + W_i^{12} \\ * & -2X_i + W_i^{22} \end{bmatrix}.
$$

Replacing $\tilde{A}_i, \tilde{B}_i, \tilde{C}_i, \bar{E}_i, \bar{H}_i$ in (8.34) by their expressions in (8.3), it follows that

$$
J_i^T G_i \tilde{A}_i J_i = \begin{bmatrix} A_i Y_i + B_i \hat{C}_i Y_i & A_i \\ X_i A_i Y_i + (Y_i^{-1} - X_i)\hat{B}_i L_i Y_i + X_i B_i \hat{C}_i Y_i + (Y_i^{-1} - X_i)\hat{A}_i Y_i & X_i A_i + (Y_i^{-1} - X_i)\hat{B}_i L_i \end{bmatrix},
$$

$$
J_i^T G_i \tilde{B}_i = \begin{bmatrix} B_{i\omega} \\ X_i B_{i\omega} + Y_i^{-1} \hat{B}_i D_{i\omega} - X_i \hat{B}_i D_{i\omega} \end{bmatrix}, J_i^T G_i \bar{E}_i = \begin{bmatrix} E_i & 0 \\ X_i E_i & 0 \end{bmatrix},
$$

$$
J_i^T \bar{H}_i = \begin{bmatrix} Y_i H_i & 0 \\ H_i & 0 \end{bmatrix}, J_i^T \tilde{C}_i^T S = \begin{bmatrix} Y_i C_i^T S + Y_i \hat{C}_i^T D_i^T S \\ C_i^T S \end{bmatrix},
$$

$$
J_i^T \tilde{C}_i^T R_-^{1/2} = \begin{bmatrix} Y_i C_i^T R_-^{1/2} + Y_i \hat{C}_i^T D_i^T R_-^{1/2} \\ C_i^T R_-^{1/2} \end{bmatrix}.
$$

Denote

$$
T_i \triangleq \hat{C}_i Y_i, V_i \triangleq (Y_i^{-1} - X_i)\hat{B}_i,
$$

$$
A_{icc} \triangleq X_i A_i Y_i + (Y_i^{-1} - X_i)\hat{B}_i L_i Y_i + X_i B_i \hat{C}_i Y_i + (Y_i^{-1} - X_i)\hat{A}_i Y_i.
$$

Then we can obtain (8.29). Similarly, conditions (8.30)–(8.31) for $N = 3$ and conditions (8.27)–(8.28) for $N = 2$ imply the dissipative condition addressed in Theorem 8.1. Therefore, Theorem 8.2 also guarantees the dissipativity of system (8.3) with the designed controller. The proof is completed. □

8.5 ILLUSTRATIVE EXAMPLES

To demonstrate the effectiveness of the proposed approach, consider the switched systems described by the following parameters:

$$A_1 = \begin{bmatrix} b & -0.5 \\ a & -0.9 \end{bmatrix}, A_2 = \begin{bmatrix} 0.9 & -a \\ -0.6 & 0.8 \end{bmatrix},$$

$$B_1 = \begin{bmatrix} 1 \\ -1 \end{bmatrix}, B_2 = \begin{bmatrix} 1 \\ 1 \end{bmatrix},$$

$$B_{1\omega} = \begin{bmatrix} 0.1 \\ 0.5 \end{bmatrix}, B_{2\omega} = \begin{bmatrix} 0.2 \\ 0.3 \end{bmatrix},$$

$$C_1 = \begin{bmatrix} 0.1 & -0.2 \end{bmatrix}, C_2 = \begin{bmatrix} 0.2 & -0.2 \end{bmatrix},$$

$$D_1 = 0.3, D_2 = 0.1,$$

$$D_{1\omega} = 0.3, D_{2\omega} = 0.1,$$

$$L_1 = \begin{bmatrix} 0.3 & 0.2 \end{bmatrix}, L_2 = \begin{bmatrix} 0.1 & 0.3 \end{bmatrix},$$

$$E_1 = \begin{bmatrix} 0.2 & 0.1 \\ 0.1 & 0.1 \end{bmatrix}, E_2 = \begin{bmatrix} 0.1 & 0.1 \\ 0.2 & 0.1 \end{bmatrix},$$

$$H_1 = \begin{bmatrix} 0.1 & 0.2 \\ 0.1 & 0.1 \end{bmatrix}, H_2 = \begin{bmatrix} 0.1 & 0.1 \\ 0.2 & 0.1 \end{bmatrix}.$$

Assign $Q = 0.85$, $S = 0.75$, and $R = -1.6$. Given $\mu = 2$, $\alpha = 0.1$, $\varepsilon_1 = 0.5$, and $\varepsilon_2 = 2$, Fig. 8.1 presents dissipative regions, which guarantee the GUAS and dissipativity of the above systems in the plane (a, b). Dissipative regions of the proposed MLF approach ($N > 1$) and the conventional approach ($N = 1$), which are commonly used under the existing ADT design framework, are drawn to make a comparison. Obviously, with the increasing of the number N, a larger dissipative region can be obtained.

As mentioned in [25], the switching must be gentler when the monotonic requirements of LF are relaxed. The ADT guaranteeing GUAS for $N = 1$, $N = 2$, $N = 3$, and $N = 4$ can be solved as $\tau_{a1}^* = 6.5788$, $\tau_{a2}^* = 13.1576$, $\tau_{a3}^* = 19.7364$, and $\tau_{a4}^* = 26.3153$, respectively.

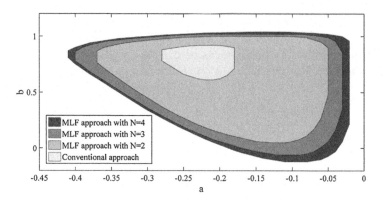

FIGURE 8.1 Comparison of the dissipative regions.

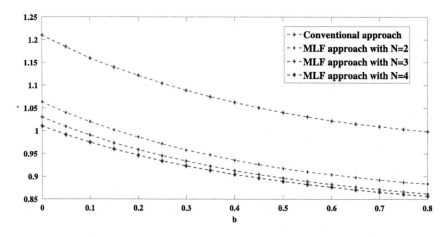

FIGURE 8.2 Comparison of the attenuation levels.

Then, a particular case, H_∞ analysis, is considered by choosing $R = -1$, $S = 0$, and $Q = \gamma^2$. To compare with the conventional LF approach ($N = 1$), we fix $a = -0.2$ and choose a set of b within the dissipative region. The comparisons of the minimal γ are shown in Fig. 8.2. It is clear that the MLF approach has better optimized H_∞ performance with the increasing of the number N.

Next, we direct our attention to verify the effectiveness of the dissipative DOF controller designed in Theorem 8.2 for $N = 2$. Assume the initial state and estimated state as $x_0 = \hat{x}_0 = [\ 0 \quad 0\ ]^T$ and the exogenous disturbance $\omega_k = 0.5e^{-0.05k}$. The simulation time is taken as 100 time units, and each unit length is taken as $T_s = 2$. The switching signal path is generated in Fig. 8.3, which satisfies the minimal ADT constraint given by $\tau_a > \tau_{a2}^* = 13.1576$. The trajectories of the state responses for system (8.1)

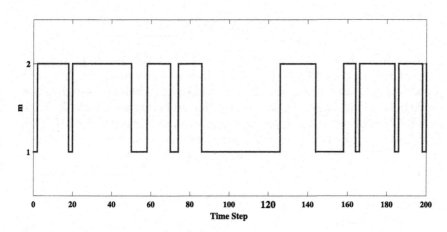

FIGURE 8.3 Switching signals.

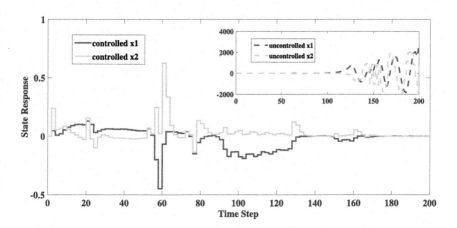

FIGURE 8.4 State trajectories.

without control (dotted lines) and under control (solid lines) are drawn in Fig. 8.4. By solving the conditions given by $N = 2$ in Theorem 8.2, the control matrices can be obtained as

$$\hat{A}_1 = \begin{bmatrix} -0.60 & -1.66 \\ 1.01 & 0.10 \end{bmatrix}, \hat{B}_1 = \begin{bmatrix} 3.53 \\ -2.70 \end{bmatrix}, \hat{C}_1 = \begin{bmatrix} -0.60 & -0.29 \end{bmatrix},$$

$$\hat{A}_2 = \begin{bmatrix} -0.08 & -0.35 \\ -1.75 & -0.60 \end{bmatrix}, \hat{B}_2 = \begin{bmatrix} 1.30 \\ 3.37 \end{bmatrix}, \hat{C}_2 = \begin{bmatrix} -0.82 & -0.44 \end{bmatrix}.$$

8.6 CONCLUSION

The robust H_∞ filtering problem for discrete-time uncertain switched systems is investigated via the N-step ahead LF approach. Such an approach can achieve better filtering performance, whereas as a cost, the restriction on the switching law is required to be more critical. To make a trade-off between filtering performance and switching law, the MDADT switching is introduced to reduce the ADT bound such that more general result, i.e., better filtering capability with smaller ADT bound, can be achieved.

References

[1] Y. Wei, J. Qiu, P. Shi, et al., Fixed-order piecewise-affine output feedback controller for fuzzy-affine-model-based nonlinear systems with time-varying delay, IEEE Transactions on Circuits and Systems I: Regular Papers 64 (4) (2017) 945–958.

[2] Y. Wei, J. Qiu, H.R. Karimi, Fuzzy-affine-model-based memory filter design of nonlinear systems with time-varying delay, IEEE Transactions on Fuzzy Systems 26 (2) (2017) 504–517.

[3] A. Hocine, M. Chadli, H.R. Karimi, A structured filter for Markovian switching systems, International Journal of Systems Science 45 (7) (2014) 1518–1527.

[4] H. Li, Y. Gao, L. Wu, et al., Fault detection for TS fuzzy time-delay systems: delta operator and input–output methods, IEEE Transactions on Cybernetics 45 (2) (2015) 229–241.

[5] Y. Gao, F. Xiao, J. Liu, et al., Distributed soft fault detection for interval type-2 fuzzy-model-based stochastic systems with wireless sensor networks, IEEE Transactions on Industrial Informatics 15 (1) (2019) 334–347.

[6] Z. Gao, Kalman filters for continuous-time fractional-order systems involving fractional-order colored noises using Tustin generating function, International Journal of Control, Automation, and Systems 16 (3) (2018) 1049–1059.

[7] S. Beidaghi, A.A. Jalali, A.K. Sedigh, New H_2 filtering for descriptor systems: singular and normal filters, International Journal of Control, Automation, and Systems 15 (1) (2017) 160–168.

[8] X. Li, J. Lam, H. Gao, et al., H_∞ and H_2 filtering for linear systems with uncertain Markov transitions, Automatica 67 (2016) 252–266.

[9] D. Du, B. Jiang, P. Shi, et al., H_∞ filtering of discrete-time switched systems with state delays via switched Lyapunov function approach, IEEE Transactions on Automatic Control 52 (8) (2007) 1520–1525.

[10] H. Liu, P. Shi, H.R. Karimi, et al., Finite-time stability and stabilisation for a class of nonlinear systems with time-varying delay, International Journal of Systems Science 47 (6) (2016) 1433–1444.

[11] E. Tian, W.K. Wong, D. Yue, et al., H_∞ filtering for discrete-time switched systems with known sojourn probabilities, IEEE Transactions on Automatic Control 60 (9) (2015) 2446–2451.

[12] K. Hu, J. Yuan, Improved robust H_∞ filtering for uncertain discrete-time switched systems, IET Control Theory & Applications 3 (3) (2009) 315–324.

[13] M. Wag, J. Qiu, M. Chadli, et al., A switched system approach to exponential stabilization of sampled-data T–S fuzzy systems with packet dropouts, IEEE Transactions on Cybernetics 46 (12) (2016) 3145–3156.

[14] P. Zhao, R. Nagamune, Switching LPV controller design under uncertain scheduling parameters, Automatica 76 (2017) 243–250.

[15] L. Zhang, E.K. Boukas, P. Shi, Exponential H_∞ filtering for uncertain discrete-time switched linear systems with average dwell time: a μ-dependent approach, International Journal of Robust and Nonlinear Control 18 (11) (2008) 1188–1207.

[16] D. Wang, P. Shi, J. Wang, et al., Delay-dependent exponential H_∞ filtering for discrete-time switched delay systems, International Journal of Robust and Nonlinear Control 22 (13) (2012) 1522–1536.

[17] L. Zhang, N. Cui, M. Liu, et al., Asynchronous filtering of discrete-time switched linear systems with average dwell time, IEEE Transactions on Circuits and Systems I: Regular Papers 58 (5) (2011) 1109–1118.

[18] J. Lian, C. Mu, P. Shi, Asynchronous H_∞ filtering for switched stochastic systems with time-varying delay, Information Sciences 224 (2013) 200–212.

[19] X. Zhao, L. Zhang, P. Shi, et al., Stability and stabilization of switched linear systems with mode-dependent average dwell time, IEEE Transactions on Automatic Control 57 (7) (2012) 1809–1815.

[20] H. Zhang, D. Xie, H. Zhang, et al., Stability analysis for discrete-time switched systems with unstable subsystems by a mode-dependent average dwell time approach, ISA Transactions 53 (4) (2014) 1081–1086.

[21] J. Cheng, H. Zhu, S. Zhong, et al., Finite-time filtering for switched linear systems with a mode-dependent average dwell time, Nonlinear Analysis: Hybrid Systems 15 (2015) 145–156.

[22] X. Zhao, H. Liu, Z. Wang, Weighted H_∞ performance analysis of switched linear systems with mode-dependent average dwell time, International Journal of Systems Science 44 (11) (2013) 2130–2139.

[23] H. Liu, X. Zhao, Finite-time H_∞ control of switched systems with mode-dependent average dwell time, Journal of the Franklin Institute 351 (3) (2014) 1301–1315.

[24] J.W. Wen, S.K. Nguang, P. Shi, X.D. Zhao, Stability and H_∞ control of discrete-time switched systems via one-step ahead Lyapunov function approach, IET Control Theory & Applications 12 (8) (2018) 1141–1147.

[25] A. Nasiri, S.K. Nguang, A. Swain, et al., Robust output feedback controller design of discrete-time Takagi–Sugeno fuzzy systems: a non-monotonic Lyapunov approach, IET Control Theory & Applications 10 (5) (2016) 545–553.

[26] S.F. Derakhshan, A. Fatehi, Non-monotonic Lyapunov functions for stability analysis and stabilization of discrete time Takagi–Sugeno fuzzy systems, International Journal of Innovative Computing, Information & Control 10 (4) (2014) 1567–1586.

[27] A. Kruszewski, R. Wang, T.M. Guerra, Nonquadratic stabilization conditions for a class of uncertain nonlinear discrete time T-S fuzzy models: a new approach, IEEE Transactions on Automatic Control 53 (2) (2008) 606–611.

[28] J. Wen, S.K. Nguang, P. Shi, et al., Robust H_∞ control of discrete-time nonhomogenous Markovian jump systems via multistep Lyapunov function approach, IEEE Transactions on Systems, Man, and Cybernetics: Systems 47 (7) (2017) 1439–1450.

[29] Y. Xie, J.W. Wen, S.K. Nguang, L. Peng, Stability, l_2-gain and robust H_∞ control for switched systems via N-step ahead Lyapunov function approach, IEEE Access 5 (1) (2017) 26400–26408.

[30] M.S. Ali, K. Meenakshi, R. Vadivel, et al., Robust H_∞ performance of discrete-time neural networks with uncertainty and time-varying delay, International Journal of Control, Automation, and Systems 16 (4) (2018) 1637–1647.

Robust H_∞ Control of Discrete-Time Nonhomogenous Markovian Jump Systems via Multistep Ahead Lyapunov Function Approach

9.1 INTRODUCTION

Technological and economical reasons motivate the development of Markovian jump linear systems (MJLSs) with an ever-increasing complexity. Many practical systems can be modeled as MJLSs such as solar boiler stations, economics, DC motor devices, RLC circuits, etc. The systems are often composed by a few interacting operation modes and a dynamical process involved with stochastic jumping, which is inherently governed by a Markov chain. The theory of stability, stabilizability, robust control, fuzzy control, and filtering problems has been intensively studied, for instance, in [1] for the discrete-time case and [2] for the continuous-time case. Some important physical applications can also been found in [3] and references therein.

As pointed out in [4–6], the scenario that the transition probabilities (TPs) of a Markov chain are time-invariant only holds when the sojourn time (the time interval between two neighboring jump instants) of each operation mode subjects to geometric distribution (in discrete-time domain). To reflect this shortcoming, the concept of nonhomogenous Markovian jump linear systems (NMJLSs) [7], where the TPs possess time-varying

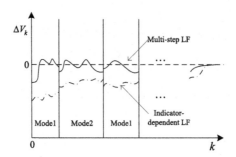

FIGURE 9.1   Time-difference of two kinds of LFs within jump mode.

character belonging to a polytope, has been introduced to describe a more general jump system model. Such an NMJLS model covers systems governed by time-invariant TPs [1], switching TPs with average dwell time [8–10], and piecewise homogenous TPs [11]. Based on the fundamental work of [7], some extended results, such as observer-based H_∞ control for NMJLS with saturated control input [12], H_∞ filtering problem of uncertain NMJLS [13], robust fault detection for discrete-time NMJLS [14], and robust filtering for nonlinear NMJLS where the indicator function is approximated by a fuzzy membership degree [15], are developed. Some related works, such as stochastic stability and stabilization of semi-Markovian jump systems [16,17] and stability analysis for MJLSs with time-delay and time-varying transition rates [18], can be found in [19–21] for a brief review.

However, the stochastic stability criteria developed for NMJLSs in the above-mentioned works are conservative because the knowledge of indicator functions is not involved in the derivation of stability criterion. Inspired by the nonmonotonic Lyapunov function (LF) [22–28], which is commonly used for the T-S fuzzy model to relax the monotonicity requirement of LFs, this chapter proposes a multistep LF that can provide certificates of stochastic stability (SS) by allowing the LF to increase during the period of several sampling time steps within the jump mode. More precisely, the time-difference of the multistep LF, i.e., ΔV_k, is allowed to be greater than zero within the jump mode (see in Fig. 9.1), whereas in the indicator LF, ΔV_k is strictly less than zero. Moreover, the multistep LF approach proposed in this chapter is much more general than the nonmonotonic LF method of two-sample variation, i.e., $V_{k+2} < V_k$, in [23–25], or N-sample variation, i.e., $V_{k+N} < V_k$, in [26–28].

Motivated by the above observations, we aim at addressing the problems of stochastic stability and stabilization for a class of discrete-time NMJLSs via a multistep LF approach. The main contribution of this chapter lies in three aspects:

- The concept of a multistep LF is proposed to reduce the conservatism of the stability criterion by allowing the LF to increase during the period of several sampling time steps within the jump mode.
- The H_∞ performance of NMJLSs is analyzed by moving the horizon from $k + N - 1$ to $k + 1$ step by step. Due to the multistep scenario, the derivation of linear matrix inequality (LMI) conditions is not trivial.
- Based on the above stability and H_∞ analysis results, a robust H_∞ controller design is proposed for NMJLSs with time-varying uncertainties.

9.2 SYSTEM DESCRIPTION AND PROBLEM FORMULATION

Consider a discrete-time NMJLS described by the following mathematical model:

$$\begin{cases} x_{k+1} = \left(A(r_k) + \grave{\Delta} A(r_k)\right) x_k + \left(B(r_k) + \grave{\Delta} B(r_k)\right) u_k + B^w(r_k)w_k, \\ z_k = C(r_k)x_k + D(r_k)u_k, \end{cases} \quad (9.1)$$

where $k \in \{0, 1, \ldots, M\}$, $x_k \in \mathbb{R}^n$ is the state vector, $u_k \in \mathbb{R}^m$ is the control input, $z_k \in \mathbb{R}^p$ is the controlled output, $w_k \in \mathbb{R}^d$ are the exogenous disturbances satisfying $\{w_k\} \in l_2[0, M]$. When $w = 0$, system (9.1) can be viewed as a *disturbance-free* NMJLS. When $u = 0$, it refers to as an *input-free* NMJLS. In the case $w = 0$ and $u = 0$, it is referred to as a *free* NMJLS.

For each possible value of $r_k = i_0$, we denote the matrices associated with the "i_0th mode" by $\mathcal{M}_{i_0}$, where $\mathcal{M}$ can be replaced by any symbols. A_{i_0}, B_{i_0}, $B_{i_0}^w$, C_{i_0}, and D_{i_0} are known constant matrices depending on the jump mode r_k. The matrices $\grave{\Delta} A_{i_0}$ and $\grave{\Delta} B_{i_0}$ represent time-varying parameter uncertainties, which are assumed to be norm bounded and can be given as

$$\begin{bmatrix} \grave{\Delta} A_{i_0} & \grave{\Delta} B_{i_0} \end{bmatrix} = E F_k \begin{bmatrix} H_{1i_0} & H_{2i_0} \end{bmatrix}, \quad (9.2)$$

where E, H_{1i_0}, and H_{2i_0} are known constant matrices characterizing the structure of the uncertainties, and F_k are unknown time-varying matrix functions with Lebesgue-measurable elements satisfying $F_k^T F_k \le I$.

A discrete-time discrete-state Markov chain r_k takes values in $\mathbb{S} = \{1, 2, \ldots, s\}$ with transition probabilities

$$\Pr(r_{k+1} = i_1 | r_k = i_0) = \pi_{i_0 i_1}(k), \quad (9.3)$$

where $\pi_{i_0 i_1}(k) \ge 0$ and $\sum_{i_1=1}^s \pi_{i_0 i_1}(k) = 1$; $\pi_{i_0 i_1}(k) \triangleq \sum_{l_0=1}^L \alpha_{l_0}(k) \pi_{i_0 i_1}^{(l_0)}$ are the entries of the TP matrix $\Pi(k)$ defined as

$$\Pi(k) \triangleq \Pi(\alpha(k)) \triangleq \sum_{l_0=1}^L \alpha_{l_0}(k)\Pi_l, \quad (9.4)$$

where $\alpha_{l_0}(k) \geq 0$, $\sum_{l_0=1}^{L} \alpha_{l_0}(k) = 1$, and Π_{l_0} $(l_0 \in \mathbb{K}_{1,L})$ are given TP matrices. We boil down all the polytopic indices of the indicator function $\alpha(k)$ over the time interval $[k, k+N]$ into a set $\mathbb{L}_{0,N} \triangleq \{l_0, l_1, \ldots, l_N\}$, where $l_0, l_1, \ldots, l_N \in \mathbb{K}_{1,L}$.

Based on the fundamental description of (9.3) and (9.4), we further define a chain of values with respect to current time jump mode r_k to jump mode r_{k+N}, which is N steps ahead. Letting $r_k = i_0$, $r_{k+1} = i_1, \ldots, r_{k+N} = i_N$, we have a series of value sets for finite-state Markov chains, i.e., $\mathbb{I}_{0,f} \triangleq \{i_0, i_1, \ldots, i_f\}$ $(f \in \mathbb{K}_{1,N})$, over the time interval $[k, k+N]$, where $i_0, i_1, \ldots, i_N \in \mathbb{S}$.

Before giving the main results, we have the following preliminaries.

Definition 9.1. The *free* jump system without uncertainties is said to be stochastically stable if

$$\lim_{M \to \infty} \mathbb{E} \left\{ \sum_{k=0}^{M} \|x_k\|^2 \right\} < \infty \tag{9.5}$$

for any finite initial conditions $x_0 \in \mathbb{R}^n$ and $r_0 \in \mathbb{S}$.

Definition 9.2. If we denote state equation (9.1) as $x_{k+1} = f(x_k)$, then the LF $\mathcal{L}(\alpha(k), r_k, x_k)$ is said to be a multistep LF if it satisfies

$$\mathcal{L}(\varphi_k) \triangleq \mathcal{L}(\alpha(k), r_k, x_k) \tag{9.6}$$

$$\triangleq \sum_{j=1}^{N} \sum_{f=j}^{N} V^f \left(\alpha(k+j-1), r_{k+j-1}, f^{j-1}(x_k) \right). \tag{9.7}$$

Remark 9.1. The superscript $j-1$ in (9.7) means calculation $j-1$ times of the function f. For example, if $N = 2$, then we have

$$\mathcal{L}(\varphi_k) = \sum_{j=1}^{2} \sum_{f=j}^{2} V^f \left(\alpha(k+j-1), r_{k+j-1}, f^{j-1}(x_k) \right),$$

which implies

$$\mathcal{L}(\varphi_k) = \sum_{f=1}^{2} V^f \left(\alpha(k), r_k, f^0(x_k) \right)$$
$$= V^1(\alpha(k), r_k, x_k) + V^2(\alpha(k), r_k, x_k) \quad (j=1)$$

and

$$\mathcal{L}(\varphi_k) = \sum_{f=2}^{2} V^f \left(\alpha(k+1), r_{k+1}, f^1(x_k) \right) = V^2(\alpha(k+1), r_{k+1}, x_{k+1}) \quad (j=2).$$

Therefore,

$$\mathcal{L}(\varphi_k) = V^1(\alpha(k), r_k, x_k) + V^2(\alpha(k), r_k, x_k) + V^2(\alpha(k+1), r_{k+1}, x_{k+1})$$

and further

$$
\begin{aligned}
\Delta\mathcal{L}(\varphi_k) &= \mathbb{E}\{\mathcal{L}(\varphi_{k+1}) \mid \mathcal{F}_k\} - \mathcal{L}(\varphi_k) \\
&= \Delta_2 V(\varphi_k) \\
&= \mathbb{E}\left\{V^2(\varphi_{k+2}) \mid \mathcal{F}_k\right\} - V^2(\varphi_k) + \mathbb{E}\left\{V^1(\varphi_{k+1}) \mid \mathcal{F}_k\right\} - V^1(\varphi_k).
\end{aligned}
$$

Remark 9.2. Multistep LF (9.6) is developed for NMJLS by improving the nonmonotonic LF introduced in Corollary 2.2 of [22]. It is clear that LF (9.6) is indicator- and mode-dependent, so that it can be employed to handle NMJLSs. Moreover, it goes to N steps ahead of the current time k, and thus it has a potential to allow the energy function increase during the finite-time interval $[k, k+N]$. Therefore, a relaxed stability condition and a better H_∞ performance could be expected. Although the difficulties lie in the derivation of the condition of H_∞ performance when the system go to future steps, we would like to develop the main results from stability condition step by step. We will demonstrate the strength of our methodology over indicator-dependent LF approach [7] and further solve the H_∞ analysis and synthesis problems in the sequel parts.

The discrete-time state feedback controller is of the form

$$u_k = K x_k = Y G^{-1} x_k. \tag{9.8}$$

Thus, the closed-loop system can be written as

$$
\begin{cases}
x_{k+1} = \tilde{A}_{i_0} x_k + B_{i_0}^w w_k, \\
z_k = \tilde{C}_{i_0} x_k,
\end{cases}
\tag{9.9}
$$

where $\tilde{A}_{i_0} = A_{i_0} + B_{i_0}K + \grave{\Delta}A_{i_0} + \grave{\Delta}B_{i_0}K$ and $\tilde{C}_{i_0} = C_{i_0} + D_{i_0}K$.

The purpose of this chapter is to design a state feedback controller of the form (9.8) such that the following two conditions are satisfied:

i) The closed-loop system (9.9) is stochastically stable.

ii) Under the zero-initial condition, the controlled output z satisfies

$$\|z\|_2^2 \leqslant \gamma^2 \|w\|_2^2, \tag{9.10}$$

where

$$\|z\|_2^2 = \mathbb{E}\left\{\sum_{k=0}^{M} z_k^T z_k\right\} < \infty \tag{9.11}$$

for any nonzero $w_k \in l_2[0, M]$ and an optimized attenuation level $\gamma > 0$.

9.3 STOCHASTIC STABILITY ANALYSIS

In this section, we first present a relaxed sufficient condition for the SS of *free* NMJLS (9.1) without uncertainties by employing the multi-step LF.

Theorem 9.1. *The free jump system (9.1) without uncertainties is stochastically stable if there exist a set of symmetric matrices* $P_{i_0}^{1,(l_0)}$, $P_{i_0,i_1}^{2,(l_1)}$, $\ldots$, $P_{\mathbb{I}_{0,f}}^{f+1,(l_f)}$, $Q_{i_1}^{1,(l_1)}$, $Q_{i_1,i_2}^{2,(l_2)}$, $\ldots$, $Q_{\mathbb{I}_{1,f}}^{f,(l_f)}$ *for all* $f \in \mathbb{K}_{1,N}$, $i_0, i_1, \ldots, i_N \in \mathbb{S}$ *and* $l_0, l_1, \ldots, l_N \in \mathbb{K}_{1,L}$ *such that*

$$\sum_{f=j}^{N} P_{\mathbb{I}_{0,f}}^{f,(l_0)} > 0 \quad \forall j \in \mathbb{K}_{1,N}, \tag{9.12a}$$

$$- Q\left(\mathbb{I}_{0,N-1}; \mathbb{L}_{0,N-1}\right) + A_{i_{N-1}}^{T} \mathcal{P}\left(\mathbb{I}_{0,N}; \mathbb{L}_{0,N}\right) A_{i_{N-1}} + \mathcal{P}\left(\mathbb{I}_{0,N-1}; \mathbb{L}_{0,N-1}\right) < 0, \tag{9.12b}$$

$$- Q\left(\mathbb{I}_{0,N-q}; \mathbb{L}_{0,N-q}\right) + A_{i_{N-q}}^{T} Q\left(\mathbb{I}_{0,N-q+1}; \mathbb{L}_{0,N-q+1}\right) A_{i_{N-q}}$$
$$+ \mathcal{P}\left(\mathbb{I}_{0,N-q}; \mathbb{L}_{0,N-q}\right) < 0 \quad \forall q \in \mathbb{K}_{2,N-1}, \tag{9.12c}$$

$$- \sum_{f=1}^{N} P_{\mathbb{I}_{0,f}}^{f,(l_0)} + A_{i_0}^{T} \left(\sum_{i_1=1}^{s} \pi_{i_0 i_1}^{(l_0)} Q_{i_1} \right) A_{i_0} < 0, \tag{9.12d}$$

where

$$\mathcal{P}\left(\mathbb{I}_{0,f}; \mathbb{L}_{0,f}\right) = \sum_{i_0=1}^{s} \pi_{i_0 i_1}^{(l_0)} \sum_{i_1=1}^{s} \pi_{i_1 i_2}^{(l_1)} \cdots \sum_{i_{f-1}=1}^{s} \pi_{i_{f-1} i_f}^{(l_{f-1})} P_{\mathbb{I}_{1,f}}^{f,(l_f)}, \tag{9.13}$$

$$\mathcal{Q}\left(\mathbb{I}_{0,f}; \mathbb{L}_{0,f}\right) = \sum_{i_0=1}^{s} \pi_{i_0 i_1}^{(l_0)} \sum_{i_1=1}^{s} \pi_{i_1 i_2}^{(l_1)} \cdots \sum_{i_{f-1}=1}^{s} \pi_{i_{f-1} i_f}^{(l_{f-1})} Q_{\mathbb{I}_{1,f}}^{f,(l_f)}. \tag{9.14}$$

Proof. Consider the multistep LF (9.6) with

$$V^f(\varphi_k) = x_k^T P^f(\varphi_k) x_k = x_k^T \left(\sum_{l_0=1}^{L} \alpha_{l_0}(k) P_{\mathbb{I}_{0,f}}^{f,(l_0)} \right) x_k$$

for $f, j \in \mathbb{K}_{1,N}$ and denote

$$\Omega_{0,f} = \sum_{l_0=1}^{L} \alpha_{l_0}(k) \sum_{l_1=1}^{L} \alpha_{l_1}(k+1) \cdots \sum_{l_f=1}^{L} \alpha_{l_f}(k+f), \quad f \in \mathbb{K}_{1,N}. \tag{9.15}$$

Condition (9.12a) implies

$$
\begin{cases}
\sum_{f=j}^{N} V^f(\varphi_k) \text{ radially unbounded for } j \in \mathbb{K}_{1,N}, \\
\sum_{f=j}^{N} V^f(\varphi_k) > 0, \forall x_k \neq 0 \text{ for } j \in \mathbb{K}_{1,N}, \\
\sum_{f=1}^{N} f V^f(\varphi_0) = 0.
\end{cases}
\tag{9.16}
$$

By virtue of Definition 9.2, we develop the following steps to prove that

$$
\begin{aligned}
\Delta \mathcal{L}(\varphi_k) &= \mathbb{E}\{\mathcal{L}(\varphi_{k+1}) \mid \mathcal{F}_k\} - \mathcal{L}(\varphi_k) \\
&= \Delta_N V(\varphi_k) \\
&= \sum_{f=1}^{N} \left(\mathbb{E}\left\{ V^f(\varphi_{k+f}) \mid \mathcal{F}_k \right\} - V^f(\varphi_k) \right) \\
&= \mathbb{E}\left\{ V^N(\varphi_{k+N}) \mid \mathcal{F}_k \right\} - V^N(\varphi_k) + \cdots + \mathbb{E}\left\{ V^1(\varphi_{k+1}) \mid \mathcal{F}_k \right\} \\
&\quad - V^1(\varphi_k) < 0
\end{aligned}
\tag{9.17}
$$

establishes the SS of free jump system (9.1) without uncertainties.

Step 1: For $f, j \in \mathbb{K}_{1,N}$, let

$$
\left(V^f(\varphi_k) \right)' \triangleq x_k^T Q^f(\varphi_k) x_k \triangleq x_k^T \left(\sum_{l_0=1}^{L} \alpha_{l_0}(k) Q_{\mathbb{I}_{0,f}}^{f,(l_0)} \right) x_k.
\tag{9.18}
$$

Adding and subtracting $\left(V^{N-1}(\varphi_{k+N-1}) \right)' = x_{k+N-1}^T \, \Omega_{0,N-1} \, \mathcal{Q} \, (\mathbb{I}_{0,N-1}; \mathbb{L}_{0,N-1}) \, x_{k+N-1}$ to and from (9.17) and bearing in mind that $x_{k+N} = A_{i_{N-1}} x_{k+N-1}$, we have

$$
\begin{aligned}
& \Delta_N V(\varphi_k) \\
=\ & x_{k+N-1}^T \Big(A_{i_{N-1}}^T \Omega_{0,N} \mathcal{P} \left(\mathbb{I}_{0,N}; \mathbb{L}_{0,N} \right) A_{i_{N-1}} + \Omega_{0,N-1} \mathcal{P} \left(\mathbb{I}_{0,N-1}; \mathbb{L}_{0,N-1} \right) \\
& - \Omega_{0,N-1} \mathcal{Q} \left(\mathbb{I}_{0,N-1}; \mathbb{L}_{0,N-1} \right) \Big) x_{k+N-1} \\
& + x_{k+N-1}^T \Omega_{0,N-1} \mathcal{Q} \left(\mathbb{I}_{0,N-1}; \mathbb{L}_{0,N-1} \right) x_{k+N-1} \\
& + x_{k+f}^T \left(\sum_{f=1}^{N-2} \Omega_{0,f} \mathcal{P} \left(\mathbb{I}_{0,f}; \mathbb{L}_{0,f} \right) \right) x_{k+f} \\
& - x_k^T \left(\sum_{l_0=1}^{L} \alpha_{l_0}(k) \sum_{f=1}^{N} P_{\mathbb{I}_{0,f}}^{f,(l_0)} \right) x_k.
\end{aligned}
\tag{9.19}
$$

If (9.12b) holds, then (9.19) leads to

$$\Delta_N V(\varphi_k) \quad < \quad x_{k+N-1}^T \Omega_{0,N-1} \mathcal{Q} \left(\mathbb{I}_{0,N-1}; \mathbb{L}_{0,N-1} \right) x_{k+N-1}$$
$$+ x_{k+f}^T \left(\sum_{f=1}^{N-2} \Omega_{0,f} \mathcal{P} \left(\mathbb{I}_{0,f}; \mathbb{L}_{0,f} \right) \right) x_{k+f}$$
$$- x_k^T \left(\sum_{l_0=1}^{L} \alpha_{l_0}(k) \sum_{f=1}^{N} P_{\mathbb{I}_{0,f}}^{f,(l_0)} \right) x_k. \tag{9.20}$$

Step 2: Adding and subtracting $\left(V^{N-2}(\varphi_{k+N-2}) \right)' = x_{k+N-2}^T \Omega_{0,N-2} \mathcal{Q} \left(\mathbb{I}_{0,N-2}; \mathbb{L}_{0,N-2} \right) x_{k+N-2}$ to and from (9.20) and bearing in mind that $x_{k+N-1} = A_{i_{N-2}} x_{k+N-2}$, we have

$$\Delta_N V(\varphi_k) \quad = \quad x_{k+N-2}^T \left(A_{i_{N-2}}^T \Omega_{0,N-1} \mathcal{Q} \left(\mathbb{I}_{0,N-1}; \mathbb{L}_{0,N-1} \right) A_{i_{N-2}} \right.$$
$$+ \Omega_{0,N-2} \mathcal{P} \left(\mathbb{I}_{0,N-2}; \mathbb{L}_{0,N-2} \right)$$
$$\left. - \Omega_{0,N-2} \mathcal{Q} \left(\mathbb{I}_{0,N-2}; \mathbb{L}_{0,N-2} \right) \right) x_{k+N-2}$$
$$+ x_{k+N-2}^T \Omega_{0,N-2} \mathcal{Q} \left(\mathbb{I}_{0,N-2}; \mathbb{L}_{0,N-2} \right) x_{k+N-2}$$
$$+ x_{k+f}^T \left(\sum_{f=1}^{N-3} \Omega_{0,f} \mathcal{P} \left(\mathbb{I}_{0,f}; \mathbb{L}_{0,f} \right) \right) x_{k+f}$$
$$- x_k^T \left(\sum_{l_0=1}^{L} \alpha_{l_0}(k) \sum_{f=1}^{N} P_{\mathbb{I}_{0,f}}^{f,(l_0)} \right) x_k. \tag{9.21}$$

If (9.12c) holds for $q = 2$, then (9.21) results in

$$\Delta_N V(\varphi_k) \quad < \quad x_{k+N-2}^T \Omega_{0,N-2} \mathcal{Q} \left(\mathbb{I}_{0,N-2}; \mathbb{L}_{0,N-2} \right) x_{k+N-2}$$
$$+ x_{k+f}^T \left(\sum_{f=1}^{N-3} \Omega_{0,f} \mathcal{P} \left(\mathbb{I}_{0,f}; \mathbb{L}_{0,f} \right) \right) x_{k+f}$$
$$- x_k^T \left(\sum_{l_0=1}^{L} \alpha_{l_0}(k) \sum_{f=1}^{N} P_{\mathbb{I}_{0,f}}^{f,(l_0)} \right) x_k. \tag{9.22}$$

Repeating the above procedure $N - 3$ times yields

$$\Delta_N V(\varphi_k) \quad < \quad x_{k+1}^T \Omega_{0,1} \mathcal{Q} \left(\mathbb{I}_{0,1}; \mathbb{L}_{0,1} \right) x_{k+1} - x_k^T \left(\sum_{l_0=1}^{L} \alpha_{l_0}(k) \sum_{f=1}^{N} P_{\mathbb{I}_{0,f}}^{f,(l_0)} \right) x_k$$

$$= x_k^T \left(A_{i_0}^T \left(\sum_{l_0=1}^{L} \alpha_{l_0}(k) \sum_{l_1=1}^{L} \alpha_{l_1}(k+1) \times \sum_{i_1=1}^{s} \pi_{i_0 i_1}^{(l_0)} Q_{i_1} \right) \right.$$

$$\left. \times A_{i_0} \sum_{l_0=1}^{L} \alpha_{l_0}(k) \sum_{f=1}^{N} P_{\mathbb{I}_{0,f}}^{f,(l_0)} \right) x_k$$

$$= x_k^T \Lambda_{\mathbb{I}_{0,f}}(k) x_k. \tag{9.23}$$

Step 3: By observing (9.23) and condition (9.12d) we obtain

$$\Delta_N V(\varphi_k) \le -\beta_{\mathbb{I}_{0,f}} x_k^T x_k \le -\beta x_k^T x_k, \tag{9.24}$$

where $\beta_{\mathbb{I}_{0,f}} = \min_k \left\{ \lambda_{\min} \left(-\Lambda_{\mathbb{I}_{0,f}}(k) \right) \right\}$ and $\beta = \min_{\mathbb{I}_{0,f} \in \mathbb{S}} \beta_{\mathbb{I}_{0,f}}$.

From (9.17) and (9.24) we have that, for any $M \ge 1$,

$$\mathbb{E} \left\{ \sum_{k=0}^{M} \Delta_N V(\varphi_k) \mid \mathcal{F}_0 \right\} = \mathbb{E} \{ \mathcal{L}(\varphi_{M+1}) \mid \mathcal{F}_0 \} - \mathcal{L}(\varphi_0)$$

$$\le -\beta \mathbb{E} \left\{ \sum_{k=0}^{M} \|x_k\|^2 \mid \mathcal{F}_0 \right\}, \tag{9.25}$$

yielding, for any $M \ge 1$,

$$\mathbb{E} \left\{ \sum_{k=0}^{M} \|x_k\|^2 \mid \mathcal{F}_0 \right\} \le \beta^{-1} \left(\mathcal{L}(\varphi_0) - \mathbb{E} \{ \mathcal{L}(\varphi_{M+1}) \mid \mathcal{F}_0 \} \right) \le \beta^{-1} \left(\mathcal{L}(\varphi_0) \right) \tag{9.26}$$

and further implying

$$\lim_{M \to \infty} \mathbb{E} \left\{ \sum_{k=0}^{M} \|x_k\|^2 \mid \mathcal{F}_0 \right\} \le \beta^{-1} \left(\mathcal{L}(\varphi_0) \right) < \infty. \tag{9.27}$$

This implies, by Definition 9.1, that free jump system (9.1) without uncertainties is stochastically stable. $\square$

Remark 9.3. An alternative way to establish Theorem 9.1 is to consider multistep LF (9.6) by replacing (9.7) by

$$V^f(\varphi_k) \triangleq x_k^T \left(\sum_{l_0=1}^{L} \alpha_{l_0}(k) P_{if}^{f,(l_0)} \right) x_k \quad \forall f, j \in \mathbb{K}_{1,N}$$

and replacing $\sum_{i_0=1}^{s} \pi_{i_0 i_1}^{(l_0)} \sum_{i_1=1}^{s} \pi_{i_1 i_2}^{(l_1)} \cdots \sum_{i_{f-1}=1}^{s} \pi_{i_{f-1} i_f}^{(l_{f-1})}$ in (9.13) and (9.14) by f-step TPs $\sum_{i_f=1}^{s} \pi_{i_0 i_f}^{(f),(l_0)}$. Following the proof of Theorem 9.1, sufficient conditions presented by LMIs are easy to obtain, and the computational efforts can be reduced to some extent. However, more conservatism is brought in at the same time because the jump modes between i_0 and i_f are totally ignored.

Remark 9.4. Based on the reviewing of [22], we know that two-sample variation means $\mathbb{E}\{V(\varphi_{k+2}) \mid \mathcal{F}_k\} - V(\varphi_k) < 0$ (similarly as in [23–25]) and N-sample variation indicates $\mathbb{E}\{V(\varphi_{k+N}) \mid \mathcal{F}_k\} - V(\varphi_k) < 0$ (similarly as in [26–28]). By observing inequality (9.17) we can conclude that if the time difference $\Delta\mathcal{L}(\varphi_k) < 0$, it allows $\mathbb{E}\{V^N(\varphi_{k+N}) \mid \mathcal{F}_k\} - V^N(\varphi_k) < 0$, or $\mathbb{E}\{V^2(\varphi_{k+2}) \mid \mathcal{F}_k\} - V^2(\varphi_k) < 0$, or any items of

$$\mathbb{E}\left\{V^f(\varphi_{k+f}) \mid \mathcal{F}_k\right\} - V^f(\varphi_k) < 0 \quad \forall f \in [1,2) \cup [3,4,\ldots,N)$$

less than zero. As a consequence, it is quite clear that $\Delta\mathcal{L}(\varphi_k) < 0$ covers $\mathbb{E}\{V(\varphi_{k+N}) \mid \mathcal{F}_k\} - V(\varphi_k) < 0$ and $\mathbb{E}\{V(\varphi_{k+2}) \mid \mathcal{F}_k\} - V(\varphi_k) < 0$ as two particular cases. Therefore, a more general approach is given in this study, and it also has great potentials to be extended to multimodel systems, such as stochastic multiagent systems [29], time-varying systems subject to quantization [30], genetic regulatory networks [31], fuzzy systems [32] with Markovian jumps, etc. However, a trade-off between the performance and computational effort must be carefully considered.

9.4 H_∞ PERFORMANCE ANALYSIS

In this section, we consider the H_∞ performance analysis for discrete-time NMJLS with exogenous disturbances under the multistep LF framework. Sufficient conditions for the γ disturbance attenuation property are given in terms of the feasibility of some coupled LMIs.

Theorem 9.2. *The input-free NMJLS (9.1) without uncertainties possesses the γ-disturbance attenuation level in (9.10) for all $w \in l_2[0, M]$, $w \neq 0$, if there exist a set of symmetric matrices $P_{i_0}^{1,(l_0)}$, $P_{i_0,i_1}^{2,(l_1)}, \ldots, P_{\mathbb{I}_{0,f}}^{f+1,(l_f)}$, $Q_{i_1}^{1,(l_1)}$, $Q_{i_1,i_2}^{2,(l_2)}, \ldots, Q_{\mathbb{I}_{1,f}}^{f,(l_f)}$ for all $f \in \mathbb{K}_{1,N}$, $i_0, i_1, \ldots, i_N \in \mathbb{S}$ and $l_0, l_1, \ldots, l_N \in \mathbb{K}_{1,L}$ such that*

$$\sum_{f=j}^{N} P_{\mathbb{I}_{0,f}}^{f,(l_0)} > 0 \quad \forall j \in \mathbb{K}_{1,N}, \tag{9.28a}$$

$$\Xi_1 = \begin{bmatrix} \Xi_1^{11} & \Xi_1^{12} & C_{i_{N-1}}^T \\ * & \Xi_1^{22} & 0_{dp} \\ * & * & -\gamma I_p \end{bmatrix} < 0, \tag{9.28b}$$

$$\Xi_q = \begin{bmatrix} \Xi_q^{11} & \Xi_q^{12} & C_{i_q}^T \\ * & \Xi_q^{22} & 0_{dp} \\ * & * & -\gamma I_p \end{bmatrix} < 0 \quad \forall q \in \mathbb{K}_{2,N-1}, \qquad (9.28c)$$

$$\Xi_N = \begin{bmatrix} \Xi_N^{11} & \Xi_1^{12} & C_{i_0}^T \\ * & \Xi_N^{22} & 0_{dp} \\ * & * & -\gamma I_p \end{bmatrix} < 0, \qquad (9.28d)$$

where

$$\Xi_1^{11} = -\mathcal{Q}\left(\mathbb{I}_{0,N-1}; \mathbb{L}_{0,N-1}\right) + A_{i_{N-1}}^T \mathcal{P}\left(\mathbb{I}_{0,N}; \mathbb{L}_{0,N}\right) A_{i_{N-1}}$$
$$+ \mathcal{P}\left(\mathbb{I}_{0,N-1}; \mathbb{L}_{0,N-1}\right),$$

$$\Xi_1^{12} = A_{i_{N-1}}^T \mathcal{P}\left(\mathbb{I}_{0,N}; \mathbb{L}_{0,N}\right) B_{i_{N-1}}^w,$$

$$\Xi_1^{22} = -\gamma I_d + \left(B_{i_{N-1}}^w\right)^T \mathcal{P}\left(\mathbb{I}_{0,N}; \mathbb{L}_{0,N}\right) B_{i_{N-1}}^w,$$

$$\Xi_q^{11} = -\mathcal{Q}\left(\mathbb{I}_{0,N-q}; \mathbb{L}_{0,N-q}\right) + A_{i_{N-q}}^T \mathcal{Q}\left(\mathbb{I}_{0,N-q+1}; \mathbb{L}_{0,N-q+1}\right) A_{i_{N-q}}$$
$$+ \mathcal{P}\left(\mathbb{I}_{0,N-q}; \mathbb{L}_{0,N-q}\right),$$

$$\Xi_q^{12} = A_{i_{N-q}}^T \mathcal{Q}\left(\mathbb{I}_{0,N-q+1}; \mathbb{L}_{0,N-q+1}\right) B_{i_{N-q}}^w,$$

$$\Xi_q^{22} = -\gamma I_d + \left(B_{i_{N-q}}^w\right)^T \mathcal{Q}\left(\mathbb{I}_{0,N-q+1}; \mathbb{L}_{0,N-q+1}\right) B_{i_{N-q}}^w,$$

$$\Xi_N^{11} = -\sum_{f=1}^{N} P_{\mathbb{I}_{0,f}}^{f,(l_0)} + A_{i_0}^T \left(\sum_{i_1=1}^{s} \pi_{i_0 i_1}^{(l_0)} Q_{i_1}\right) A_{i_0},$$

$$\Xi_N^{12} = A_{i_0}^T \left(\sum_{i_1=1}^{s} \pi_{i_0 i_1}^{(l_0)} Q_{i_1}\right) B_{i_0}^w,$$

$$\Xi_N^{22} = -\gamma I_d + \left(B_{i_0}^w\right)^T \left(\sum_{i_1=1}^{s} \pi_{i_0 i_1}^{(l_0)} Q_{i_1}\right) B_{i_0}^w.$$

Proof. It is clear that LMIs (9.28a)–(9.28d) incorporates sufficient conditions presented in Theorem 9.1, and therefore Theorem 9.2 implies the SS of the input-free NMJLS (9.1) without uncertainties. Next, we direct our attention to the H_∞ performance analysis.

Step 1: Define the new vector $\zeta_k \triangleq \begin{bmatrix} x_k^T & w_k^T \end{bmatrix}^T$. Adding and subtracting $x_{k+N-1}^T \Omega_{0,N-1} \mathcal{Q}\left(\mathbb{I}_{0,N-1}; \mathbb{L}_{0,N-1}\right) x_{k+N-1} + \gamma^{-1} z_{k+N-1}^T z_{k+N-1} - \gamma w_{k+N-1}^T w_{k+N-1}$ to and from (9.17), we have

$$\Delta_N V(\varphi_k)$$
$$= \zeta_{k+N-1}^T \Theta_1 \zeta_{k+N-1} + x_{k+N-1}^T \Omega_{0,N-1} \mathcal{Q}\left(\mathbb{I}_{0,N-1}; \mathbb{L}_{0,N-1}\right) x_{k+N-1}$$

$$+x_{k+f}^T \left(\sum_{f=1}^{N-2} \Omega_{0,f} \mathcal{P} \left(\mathbb{I}_{0,f}; \mathbb{L}_{0,f} \right) \right) x_{k+f}$$

$$-x_k^T \left(\sum_{l_0=1}^{L} \alpha_{l_0}(k) \sum_{f=1}^{N} P_{\mathbb{I}_{0,f}}^{f,(l_0)} \right) x_k - \gamma^{-1} z_{k+N-1}^T z_{k+N-1}$$

$$+\gamma w_{k+N-1}^T w_{k+N-1}, \tag{9.29}$$

where

$$\Theta_1 = \begin{bmatrix} \Theta_1^{11} & \Xi_1^{12} \\ * & \Xi_1^{22} \end{bmatrix}, \Theta_1^{11} = \Xi_1^{11} + \gamma^{-1} C_{i_{N-1}}^T C_{i_{N-1}}.$$

Inequality (9.28b) implies $\zeta_{k+N-1}^T \Theta_1 \zeta_{k+N-1} < 0$, therefore

$$\Delta_N V(\varphi_k) < x_{k+N-1}^T \Omega_{0,N-1} \mathcal{Q} \left(\mathbb{I}_{0,N-1}; \mathbb{L}_{0,N-1} \right) x_{k+N-1}$$

$$+x_{k+f}^T \left(\sum_{f=1}^{N-2} \Omega_{0,f} \mathcal{P} \left(\mathbb{I}_{0,f}; \mathbb{L}_{0,f} \right) \right) x_{k+f}$$

$$-x_k^T \left(\sum_{l_0=1}^{L} \alpha_{l_0}(k) \sum_{f=1}^{N} P_{\mathbb{I}_{0,f}}^{f,(l_0)} \right) x_k - \gamma^{-1} z_{k+N-1}^T z_{k+N-1}$$

$$+\gamma w_{k+N-1}^T w_{k+N-1}. \tag{9.30}$$

Step 2: Adding and subtracting $x_{k+N-2}^T \Omega_{0,N-2} \mathcal{Q} \left(\mathbb{I}_{0,N-2}; \mathbb{L}_{0,N-2} \right) x_{k+N-2}$ $+ \gamma^{-1} z_{k+N-1}^T z_{k+N-1} - \gamma w_{k+N-1}^T w_{k+N-1}$ to and from (9.30), we have

$$\Delta_N V(\varphi_k) < \zeta_{k+N-2}^T \Theta_2 \zeta_{k+N-2}$$

$$+x_{k+N-2}^T \Omega_{0,N-2} \mathcal{Q} \left(\mathbb{I}_{0,N-2}; \mathbb{L}_{0,N-2} \right) x_{k+N-2}$$

$$+x_{k+f}^T \left(\sum_{f=1}^{N-3} \Omega_{0,f} \mathcal{P} \left(\mathbb{I}_{0,f}; \mathbb{L}_{0,f} \right) \right) x_{k+f}$$

$$-x_k^T \left(\sum_{l_0=1}^{L} \alpha_{l_0}(k) \sum_{f=1}^{N} P_{\mathbb{I}_{0,f}}^{f,(l_0)} \right) x_k$$

$$- \sum_{f=N-2}^{N-1} \left(\gamma^{-1} z_{k+f}^T z_{k+f} - \gamma w_{k+f}^T w_{k+f} \right) \tag{9.31}$$

where

$$
\Theta_2 = \begin{bmatrix} \Theta_2^{11} & \Xi_2^{12} \\ * & \Xi_2^{22} \end{bmatrix}, \Theta_2^{11} = \Xi_2^{11} + \gamma^{-1} C_{i_{N-2}}^T C_{i_{N-2}}.
$$

If (9.28c) holds for $q = 2$, then we obtain

$$
\begin{aligned}
\Delta_N V(\varphi_k) \quad < \quad & x_{k+N-2}^T \Omega_{0,N-2} \mathcal{Q}\left(\mathbb{I}_{0,N-2}; \mathbb{L}_{0,N-2}\right) x_{k+N-2} \\
& + x_{k+f}^T \left(\sum_{f=1}^{N-3} \Omega_{0,f} \mathcal{P}\left(\mathbb{I}_{0,f}; \mathbb{L}_{0,f}\right) \right) x_{k+f} \\
& - x_k^T \left(\sum_{l_0=1}^{L} \alpha_{l_0}(k) \sum_{f=1}^{N} P_{\mathbb{I}_{0,f}}^{f,(l_0)} \right) x_k \\
& - \sum_{f=N-2}^{N-1} \left(\gamma^{-1} z_{k+f}^T z_{k+f} - \gamma w_{k+f}^T w_{k+f} \right). \quad (9.32)
\end{aligned}
$$

Repeating the above procedure $N - 3$ times yields

$$
\begin{aligned}
& \Delta_N V(\varphi_k) \\
< \quad & x_{k+1}^T \Omega_{0,1} \mathcal{Q}\left(\mathbb{I}_{0,1}; \mathbb{L}_{0,1}\right) x_{k+1} - x_k^T \left(\sum_{l_0=1}^{L} \alpha_{l_0}(k) \sum_{f=1}^{N} P_{\mathbb{I}_{0,f}}^{f,(l_0)} \right) x_k \\
& - \sum_{f=1}^{N-1} \left(\gamma^{-1} z_{k+f}^T z_{k+f} - \gamma w_{k+f}^T w_{k+f} \right) + \gamma^{-1} z_k^T z_k - \gamma w_k^T w_k \\
& - \left(\gamma^{-1} z_k^T z_k - \gamma w_k^T w_k \right) \\
= \quad & \zeta_k^T \Theta_N \zeta_k - \sum_{f=0}^{N-1} \left(\gamma^{-1} z_{k+f}^T z_{k+f} - \gamma w_{k+f}^T w_{k+f} \right), \quad (9.33)
\end{aligned}
$$

where

$$
\Theta_N = \begin{bmatrix} \Theta_N^{11} & \Xi_N^{12} \\ * & \Xi_N^{22} \end{bmatrix}, \Theta_N^{11} = \Xi_N^{11} + \gamma^{-1} C_{i_0}^T C_{i_0}.
$$

Step 3: In the following, we assume the zero initial conditions, that is, $x_0 = x_{-1} = \cdots x_{-N} = 0$. Define $\vartheta_{k+f} \triangleq \gamma^{-1} z_{k+f}^T z_{k+f} - \gamma w_{k+f}^T w_{k+f}$ and

$$
J_M \triangleq \|z\|_2^2 - \gamma^2 \|w\|_2^2 \triangleq \mathbb{E}\left\{ \sum_{k=0}^{M} \left(z_k^T z_k - \gamma^2 w_k^T w_k \right) \right\}. \quad (9.34)
$$

Condition (9.28d) implies

$$\Delta_N V(\varphi_k) + \sum_{f=0}^{N-1} \vartheta_{k+f} < 0,$$

which further results in

$$\sum_{k=-N}^{M} \left(\sum_{f=1}^{N} \Delta_N V(\varphi_k) + \sum_{f=0}^{N-1} \vartheta_{k+f} \right) < 0$$

$$\Rightarrow \sum_{k=-N}^{M} \left(\sum_{f=1}^{N} \left(\mathbb{E}\left\{ V^f(\varphi_{k+f}) \mid \mathcal{F}_k \right\} - V^f(\varphi_k) \right) + \sum_{f=0}^{N-1} \vartheta_{k+f} \right) < 0$$

$$\Rightarrow \mathbb{E}\left\{ V^N(\varphi_{M+N}) \mid \mathcal{F}_M \right\} + \mathbb{E}\left\{ V^N(\varphi_{M+N-1}) \mid \mathcal{F}_{M-1} \right\}$$

$$\underbrace{+ \cdots + \mathbb{E}\left\{ V^N(\varphi_{M+1}) \mid \mathcal{F}_{-N+M+1} \right\}}_{N} + \cdots\cdots$$

$$\underbrace{+ \mathbb{E}\left\{ V^2(\varphi_{M+2}) \mid \mathcal{F}_M \right\} + \mathbb{E}\left\{ V^2(\varphi_{M+1}) \mid \mathcal{F}_{M-1} \right\}}_{2} + \underbrace{\mathbb{E}\left\{ V^1(\varphi_{M+1}) \mid \mathcal{F}_M \right\}}_{1}$$

$$- \mathbb{E}\left\{ V^N(\varphi_{-N}) \mid \mathcal{F}_{-N} \right\} - \mathbb{E}\left\{ V^N(\varphi_{-N+1}) \mid \mathcal{F}_{-N+1} \right\}$$

$$\underbrace{- \cdots - \mathbb{E}\left\{ V^N(\varphi_{-1}) \mid \mathcal{F}_{-1} \right\}}_{N} - \cdots\cdots$$

$$\underbrace{- \mathbb{E}\left\{ V^2(\varphi_{-N}) \mid \mathcal{F}_{-N} \right\} - \mathbb{E}\left\{ V^2(\varphi_{-N+1}) \mid \mathcal{F}_{-N+1} \right\}}_{2}$$

$$\underbrace{- \mathbb{E}\left\{ V^1(\varphi_{-N}) \mid \mathcal{F}_{-N} \right\}}_{1} + \sum_{k=-N}^{M} (\vartheta_k + \vartheta_{k+1} + \cdots + \vartheta_{k+N-1}) < 0.$$

As $M \to \infty$, all the negative items tend to zero, and we obtain

$$\frac{N(N+1)}{2} V(\varphi_\infty) + N J_\infty < 0,$$

which yields $J_\infty < 0$. Therefore, the dissipativity inequality (9.10) holds for $M > 0$. This completes the proof. $\qquad\square$

9.5 ROBUST H_∞ DISTURBANCE ATTENUATION

In this section, we consider the robust H_∞ disturbance attenuation for discrete-time NMJLS (9.1) using a multistep LF approach.

Theorem 9.3. *For NMJLS (9.1), there exists a feedback control (9.8) such that the closed-loop system is stochastically stable and possesses a disturbance attenuation level γ for all the admissible uncertainties if there exist a set of symmetric matrices $P_{i_0}^{1,(l_0)}$, $P_{i_0,i_1}^{2,(l_1)}, \ldots, P_{\mathbb{I}_{0,f}}^{f+1,(l_f)}$, $Q_{i_1}^{1,(l_1)}$, $Q_{i_1,i_2}^{2,(l_2)}, \ldots, Q_{\mathbb{I}_{1,f}}^{f,(l_f)}$ for all $f \in \mathbb{K}_{1,N}$, $i_0, i_1, \ldots, i_N \in \mathbb{S}, l_0, l_1, \ldots, l_N \in \mathbb{K}_{1,L}$, and matrices Y and G such that*

$$\sum_{f=j}^{N} P_{\mathbb{I}_{0,f}}^{f,(l_0)} > 0 \quad \forall j \in \mathbb{K}_{1,N}, \tag{9.35a}$$

$$\Upsilon_1 = \begin{bmatrix} \begin{pmatrix} -\mathcal{Q}\left(\mathbb{I}_{0,N-1};\mathbb{L}_{0,N-1}\right) \\ +\mathcal{P}\left(\mathbb{I}_{0,N-1};\mathbb{L}_{0,N-1}\right) \end{pmatrix} & 0_{nd} & \bar{C}_{i_{N-1}}^T & \bar{A}_{i_{N-1}}^T & \bar{H}_{i_{N-1}}^T \\ * & -\gamma I_d & 0_{dp} & \left(B_{i_{N-1}}^w\right)^T & 0_{dn} \\ * & * & -\gamma I_p & 0_{pn} & 0_{pn} \\ * & * & * & \begin{pmatrix} -G - G^T + \varepsilon E E^T \\ +\mathcal{P}\left(\mathbb{I}_{0,N};\mathbb{L}_{0,N}\right) \end{pmatrix} & 0_{nn} \\ * & * & * & * & -\varepsilon I_n \end{bmatrix} < 0, \tag{9.35b}$$

$$\Upsilon_q = \begin{bmatrix} \begin{pmatrix} -\mathcal{Q}\left(\mathbb{I}_{0,N-q};\mathbb{L}_{0,N-q}\right) \\ +\mathcal{P}\left(\mathbb{I}_{0,N-q};\mathbb{L}_{0,N-q}\right) \end{pmatrix} & 0_{nd} & \bar{C}_{i_{N-q}}^T & \bar{A}_{i_{N-q}}^T & \bar{H}_{i_{N-q}}^T \\ * & -\gamma I_d & 0_{dp} & \left(B_{i_{N-q}}^w\right)^T & 0_{dn} \\ * & * & -\gamma I_p & 0_{pn} & 0_{pn} \\ * & * & * & \begin{pmatrix} -G - G^T + \varepsilon E E^T \\ +\mathcal{Q}\left(\mathbb{I}_{0,N-q+1};\mathbb{L}_{0,N-q+1}\right) \end{pmatrix} & 0_{nn} \\ * & * & * & * & -\varepsilon I_n \end{bmatrix} < 0$$

$$\forall q \in \mathbb{K}_{2,N-1}, \tag{9.35c}$$

$$\Upsilon_N = \begin{bmatrix} -\sum_{f=1}^{N} P_{\mathbb{I}_{0,f}}^{f,(l_0)} & 0_{nd} & \bar{C}_{i_0}^T & \bar{A}_{i_0}^T & \bar{H}_{i_0}^T \\ * & -\gamma I_d & 0_{dp} & \left(B_{i_0}^w\right)^T & 0_{dn} \\ * & * & -\gamma I_p & 0_{pn} & 0_{pn} \\ * & * & * & \begin{pmatrix} -G - G^T + \varepsilon E E^T \\ +\sum_{i_1=1}^{s} \pi_{i_0 i_1}^{(l_0)} Q_{i_1} \end{pmatrix} & 0_{nn} \\ * & * & * & * & -\varepsilon I_n \end{bmatrix} < 0, \tag{9.35d}$$

where

$$\bar{A}_i = A_i G + B_i Y, \bar{H}_i = H_{1i} G + H_{2i} Y,$$
$$\bar{C}_i = C_i G + D_i Y, \tilde{\Delta}_i = E F_k \bar{H}_i \quad \forall i \in \mathbb{I}_{0,N}.$$

Proof. For $f, j \in \mathbb{K}_{1,N}$, let $\tilde{A}_i \triangleq \left(\bar{A}_i + \bar{\bar{\Delta}}_i\right) G^{-1}$, $\tilde{C}_i \triangleq \bar{C}_i G^{-1}$ $(i \in \mathbb{I}_{0,N})$,

$$\bar{V}^f(\varphi_k) \triangleq x_k^T \left(\sum_{l_0=1}^{L} \alpha_{l_0}(k) \bar{P}_{\mathbb{I}_{0,f}}^{f,(l_0)} \right) x_k \triangleq x_k^T \left(\sum_{l_0=1}^{L} \alpha_{l_0}(k) G^{-T} P_{\mathbb{I}_{0,f}}^{f,(l_0)} G^{-1} \right) x_k,$$
(9.36)

$$\left(\bar{V}^f(\varphi_k) \right)' \triangleq x_k^T \left(\sum_{l_0=1}^{L} \alpha_{l_0}(k) \bar{Q}_{\mathbb{I}_{0,f}}^{f,(l_0)} \right) x_k$$

$$\triangleq x_k^T \left(\sum_{l_0=1}^{L} \alpha_{l_0}(k) G^{-T} Q_{\mathbb{I}_{0,f}}^{f,(l_0)} G^{-1} \right) x_k,$$
(9.37)

$$\bar{P}\left(\mathbb{I}_{0,f}; \mathbb{L}_{0,f}\right) \triangleq \sum_{i_0=1}^{s} \pi_{i_0 i_1}^{(l_0)} \sum_{i_1=1}^{s} \pi_{i_1 i_2}^{(l_1)} \cdots \sum_{i_{f-1}=1}^{s} \pi_{i_{f-1} i_f}^{(l_{f-1})} \bar{P}_{\mathbb{I}_{1,f}}^{f,(l_f)},$$
(9.38)

$$\bar{Q}\left(\mathbb{I}_{0,f}; \mathbb{L}_{0,f}\right) \triangleq \sum_{i_0=1}^{s} \pi_{i_0 i_1}^{(l_0)} \sum_{i_1=1}^{s} \pi_{i_1 i_2}^{(l_1)} \cdots \sum_{i_{f-1}=1}^{s} \pi_{i_{f-1} i_f}^{(l_{f-1})} \bar{Q}_{\mathbb{I}_{1,f}}^{f,(l_f)}.$$
(9.39)

Based on the results of Theorem 9.2, by replacing A_i by $\tilde{A}_i$, C_i by $\tilde{C}_i$, $\mathcal{P}\left(\mathbb{I}_{0,f}; \mathbb{L}_{0,f}\right)$ by $\bar{\mathcal{P}}\left(\mathbb{I}_{0,f}; \mathbb{L}_{0,f}\right)$, and $\mathcal{Q}\left(\mathbb{I}_{0,f}; \mathbb{L}_{0,f}\right)$ by $\bar{\mathcal{Q}}\left(\mathbb{I}_{0,f}; \mathbb{L}_{0,f}\right)$, we can prove the results presented in Theorem 9.3.

It is clear that $\Xi_1 < 0$ implies $\Theta_1 < 0$ and further $\Phi_1 < 0$, where

$$\Phi_1 = \begin{bmatrix} \Phi_1^{11} & \Phi_1^{12} \\ * & \Phi_1^{22} \end{bmatrix},$$
(9.40)

$$\Phi_1^{11} = -\bar{\mathcal{Q}}\left(\mathbb{I}_{0,N-1}; \mathbb{L}_{0,N-1}\right) + \tilde{A}_{i_{N-1}}^T \bar{\mathcal{P}}\left(\mathbb{I}_{0,N}; \mathbb{L}_{0,N}\right) \tilde{A}_{i_{N-1}}$$
$$+ \bar{\mathcal{P}}\left(\mathbb{I}_{0,N-1}; \mathbb{L}_{0,N-1}\right) + \gamma^{-1} \tilde{C}_{i_{N-1}}^T \tilde{C}_{i_{N-1}},$$

$$\Phi_1^{12} = \tilde{A}_{i_{N-1}}^T \bar{\mathcal{P}}\left(\mathbb{I}_{0,N}; \mathbb{L}_{0,N}\right) B_{i_{N-1}}^w,$$

$$\Phi_1^{22} = -\gamma I_d + \left(B_{i_{N-1}}^w \right)^T \bar{\mathcal{P}}\left(\mathbb{I}_{0,N}; \mathbb{L}_{0,N}\right) B_{i_{N-1}}^w.$$

Performing a congruence transformation to $\Phi_1 < 0$ by diag$\{G^T, I\}$ and diag$\{G, I\}$, we get $\bar{\Phi}_1^{11} < 0$, where

$$\bar{\Phi}_1 = \begin{bmatrix} \bar{\Phi}_1^{11} & \bar{\Phi}_1^{12} \\ * & \bar{\Phi}_1^{22} \end{bmatrix},$$
(9.41)

$$\bar{\Phi}_1^{11} = -\bar{\mathcal{Q}}\left(\mathbb{I}_{0,N-1}; \mathbb{L}_{0,N-1}\right) + \bar{A}_{i_{N-1}}^T \bar{\mathcal{P}}\left(\mathbb{I}_{0,N}; \mathbb{L}_{0,N}\right) \bar{A}_{i_{N-1}}$$
$$+ \bar{\mathcal{P}}\left(\mathbb{I}_{0,N-1}; \mathbb{L}_{0,N-1}\right) + \gamma^{-1}\bar{C}_{i_{N-1}}^T \bar{C}_{i_{N-1}},$$

$$\bar{\Phi}_1^{12} = \bar{A}_{i_{N-1}}^T \bar{\mathcal{P}}\left(\mathbb{I}_{0,N}; \mathbb{L}_{0,N}\right) B_{i_{N-1}}^w.$$

Noting that

$$-\left(\bar{\mathcal{P}}\left(\mathbb{I}_{0,N}; \mathbb{L}_{0,N}\right)\right)^{-1} \le -G - G^T + \mathcal{P}\left(\mathbb{I}_{0,N}; \mathbb{L}_{0,N}\right)$$

and employing the Schur complement to $\bar{\Phi}_1 < 0$, we know that $\bar{\Phi}_1 < 0$ only if

$$\tilde{\Phi}_1 = \begin{bmatrix} \begin{pmatrix} -\mathcal{Q}\left(\mathbb{I}_{0,N-1}; \mathbb{L}_{0,N-1}\right) \\ +\mathcal{P}\left(\mathbb{I}_{0,N-1}; \mathbb{L}_{0,N-1}\right) \end{pmatrix} & 0_{nd} & \bar{C}_{i_{N-1}}^T & \bar{A}_{i_{N-1}}^T + \tilde{\Delta}_{i_{N-1}}^T \\ * & -\gamma I_d & 0_{dp} & \left(B_{i_{N-1}}^w\right)^T \\ * & * & -\gamma I_p & 0_{pn} \\ * & * & * & -G - G^T + \mathcal{P}\left(\mathbb{I}_{0,N}; \mathbb{L}_{0,N}\right) \end{bmatrix} < 0.$$

If we denote

$$\tilde{E} = \begin{bmatrix} 0 & 0 & 0 & 0 \\ 0 & 0 & 0 & 0 \\ 0 & 0 & 0 & 0 \\ E & 0 & 0 & 0 \end{bmatrix}, \tilde{F}_{k+N-1} = \begin{bmatrix} F_{k+N-1} & 0 & 0 & 0 \\ 0 & 0 & 0 & 0 \\ 0 & 0 & 0 & 0 \\ 0 & 0 & 0 & 0 \end{bmatrix},$$

$$\tilde{H}_{k+N-1} = \begin{bmatrix} \bar{H}_{k+N-1} & 0 & 0 & 0 \\ 0 & 0 & 0 & 0 \\ 0 & 0 & 0 & 0 \\ 0 & 0 & 0 & 0 \end{bmatrix},$$

then $\tilde{\Phi}_1$ can be decomposed as

$$\tilde{\Phi}_1 = \tilde{\Phi}_0 + \tilde{E}\tilde{F}_{k+N-1}\tilde{H}_{k+N-1} + \left(\tilde{E}\tilde{F}_{k+N-1}\tilde{H}_{k+N-1}\right)^T,$$

where

$$\tilde{\Phi}_0 = \begin{bmatrix} \begin{pmatrix} -\mathcal{Q}\left(\mathbb{I}_{0,N-1}; \mathbb{L}_{0,N-1}\right) \\ +\mathcal{P}\left(\mathbb{I}_{0,N-1}; \mathbb{L}_{0,N-1}\right) \end{pmatrix} & 0_{nd} & \bar{C}_{i_{N-1}}^T & \bar{A}_{i_{N-1}}^T \\ * & -\gamma I_d & 0_{dp} & \left(B_{i_{N-1}}^w\right)^T \\ * & * & -\gamma I_p & 0_{pn} \\ * & * & * & -G - G^T + \mathcal{P}\left(\mathbb{I}_{0,N}; \mathbb{L}_{0,N}\right) \end{bmatrix} < 0.$$

Applying the well-known inequality [20]

$$\tilde{\Phi}_0 + \tilde{E}\tilde{F}_{k+N-1}\tilde{H}_{k+N-1} + \left(\tilde{E}\tilde{F}_{k+N-1}\tilde{H}_{k+N-1}\right)^T$$
$$\le \tilde{\Phi}_0 + \varepsilon \tilde{E}\tilde{E}^T + \varepsilon^{-1}\tilde{H}_{k+N-1}^T\tilde{H}_{k+N-1},$$

we can conclude that $\tilde{\Phi}_1 < 0$ only if

$$
\begin{bmatrix}
\begin{pmatrix} -\mathcal{Q}\left(\mathbb{I}_{0,N-1};\mathbb{L}_{0,N-1}\right) \\ +\mathcal{P}\left(\mathbb{I}_{0,N-1};\mathbb{L}_{0,N-1}\right) \\ +\varepsilon^{-1}\tilde{H}_{k+N-1}^T\tilde{H}_{k+N-1} \end{pmatrix} & 0_{nd} & \bar{C}_{i_{N-1}}^T & \tilde{A}_{i_{N-1}}^T \\
* & -\gamma I_d & 0_{dp} & \left(B_{i_{N-1}}^w\right)^T \\
* & * & -\gamma I_p & 0_{pn} \\
* & * & * & \begin{pmatrix} -G - G^T \\ +\mathcal{P}\left(\mathbb{I}_{0,N};\mathbb{L}_{0,N}\right) + \varepsilon \tilde{E}\tilde{E}^T \end{pmatrix}
\end{bmatrix} < 0, \quad (9.42)
$$

which can be directly implied by $\Upsilon_1 < 0$.

Following the same lines of the proof of (9.42), we can prove that the controller obtained by Theorem 9.3 can guarantee the SS of the closed-loop system (9.9) with γ-disturbance attenuation and all the admissible uncertainties. $\qquad\square$

9.6 NUMERICAL EXAMPLES

In the numerical computation, we solved LMIs via Matlab 2009a with Yalmip and Sedumi toolbox on a PC with Intel (R) Core (TM) i5 5200U CPU @2.20 GHz.

Example 9.1. Consider the free NMJLS (9.1) without uncertainties incorporating with two operation modes and the following representation:

$$A_1 = \begin{bmatrix} b & -0.5 \\ a & -0.9 \end{bmatrix}, \quad A_2 = \begin{bmatrix} 0.9 & -a \\ -0.6 & 0.8 \end{bmatrix}.$$

The TP matrix is assumed to be time-varying in a polytope defined by its vertices

$$\Pi_1 = \begin{bmatrix} 0.1 & 0.9 \\ 0.2 & 0.8 \end{bmatrix}, \quad \Pi_2 = \begin{bmatrix} 0.3 & 0.7 \\ 0.4 & 0.6 \end{bmatrix}.$$

The indicator functions are chosen as

$$\alpha_1(k) = \frac{1 - x_1}{2}, \quad \alpha_2(k) = \frac{1 + x_1}{2}.$$

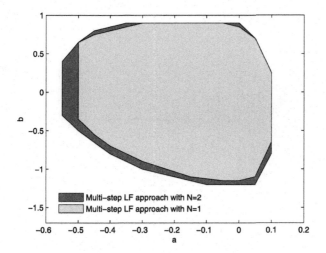

FIGURE 9.2 Stability regions using multistep LF approach.

The initial state and initial modes are taken as $x_0 = [-0.3 \quad 0.4]^T$ and $r_0 = 1$, respectively. The iterative step is taken as *length* = 30, 60, or 200. The simulation time is taken as *length* time units, and each unit length is taken as $t_s = 1$. The jumping modes path from time step 0 to time step *length* is generated randomly by employing the TP matrices Π^l and indicator function $\alpha_l(k)$.

(A) Comparison of the stability region

Standard quadratic stability conditions with common P matrix cannot justify the SS of this model whatever the pair (a, b) is. Fig. 9.2 demonstrates the stability regions by employing the multistep LF approach developed in Theorem 9.1 with $N = 1$ and $N = 2$. Note that the case $N = 1$ is the same as in the indicator-dependent LF method developed in [7].

The area marked in the light blue is the region where the solutions exist using the indicator-dependent LF method (i.e., $N = 1$). The blue area represents the region obtained from Theorem 9.1 with $N = 2$. It is clear from Fig. 9.2, that the proposed approach yields a larger stability region.

(B) Judgement of the stochastic stability

For an example, we also can take the point $(a, b) = (-0.55, 0)$ in Fig. 9.2, which is out of the light blue area but inside the blue area. We can check that the eigenvalues of the system with such a pair (a, b) are

$$\text{eig}(A_1) = \begin{bmatrix} 0.2410 \\ -1.1410 \end{bmatrix}, \text{eig}(A_2) = \begin{bmatrix} 0.85 + 0.5723i \\ 0.85 - 0.5723i \end{bmatrix},$$

which implies that each operation mode is unstable. However, state trajectories drawn in Fig. 9.3, justify the SS of the whole NMJLS.

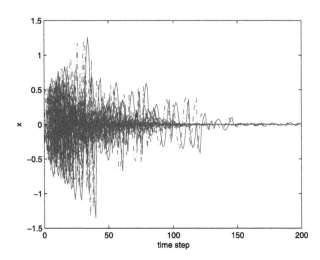

FIGURE 9.3 State trajectories of NMJLS under 30 different jumping mode paths.

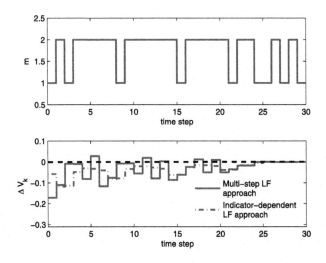

FIGURE 9.4 Comparison of the time difference of LF within the jump mode.

(C) Evolution of the time difference of LF

If we set $a = 0$ and $b = 0.4$, then Fig. 9.4 shows the evolution of the time difference of LF within the jump mode, that is, the evolution of ΔV_k. It can be seen that the time difference under indicator-dependent LF method (drawn in dash-dot line) is less than 0, which implies that the LF of each operating mode has a decreasing nature. However, the line of the multi-step LF with $N = 2$ (drawn in solid line) is larger than 0 at several time steps, which illustrates that the multistep LF is allowed to increase during

TABLE 9.1 Comparison of the number of decision variables.

Number of decision variables	General case	Example 1
Multistep LF	$\sum_{f=1}^{N-1}\left(s^f Ln(n+1)\right) + s^N L\frac{n(n+1)}{2}$	48
Indicator-dependent LF	$sL\frac{n(n+1)}{2}$	12

TABLE 9.2 Comparison of the attenuation level γ (H_∞ analysis).

a		-0.55	-0.5	-0.45	-0.4	-0.35
γ	Multistep LF	5.2025	3.9705	3.0487	2.4187	1.9897
γ	Indicator-dependent LF	Infeasible	10.4907	4.1995	2.6895	2.0066

the period of several sampling time steps and can reduce the conservatism lies in the stability criterion.

(D) Comparison of the number of decision variables

By observing Theorem 9.1 in this paper and Proposition 1 in [7] the number of decision variables can be seen in Table 9.1. Obviously, the multistep LF approach has more decision variables, and it is the cost to obtain a more flexible stability criterion.

(E) Comparison of the H_∞ performance

If we set

$$b = 0.6,\ B_1^w = \begin{bmatrix} 0.4 \\ 0.5 \end{bmatrix},\ B_2^w = \begin{bmatrix} 0.2 \\ 0.6 \end{bmatrix},$$

$$C_1 = [0.2 \quad 0.1],\ C_2 = [0.3 \quad 0.4],\ \Pi_2 = \begin{bmatrix} 0 & 1 \\ 1 & 0 \end{bmatrix}$$

and consider the input-free NMJLS (9.1) without uncertainties, with the increase of the parameter a, then the optimal attenuation level γ_{min} is computed for the cases $N = 1, 2$. Table 9.2 compares the results of multistep LF approach and indicator-dependent LF method developed in [7]. As expected, the multistep LF approach gives better optimized H_∞ performance.

Example 9.2. A modified economic model (see in Fig. 9.5) borrowed from [33] is established as an NMJLS to verify the application potential of the multistep LF approach.

Such an NMJLS contains two operation modes, depression and sustainable growth. The system state and the control law represent national income and government expenditure, respectively. The evolution of the economy is governed by a discrete-time Markov chain with time-varying transition probability (TVTP). The TVTP will be affected by the quantity of labor force, the consumption level, the productive quality, etc., from

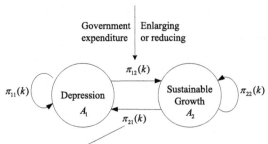

FIGURE 9.5   Scheme of an NMJLS with TVTP.

season to season or from year to year. For instance, the probability of remaining in the growth regime will increase with a great rise in the consumption. The government can regulate and control the national economy by enlarging or reducing the expenditure. Our purpose here is to show the effectiveness of the controller design via the multistep LF approach presented in Theorem 9.3. The system parameters of NMJLS (9.1) are presented as follows:

$$A_1 = \begin{bmatrix} 0 & 1 \\ -0.913 & 0.764 \end{bmatrix}, A_2 = \begin{bmatrix} -0.7 & 1 \\ 0.336 & -0.525 \end{bmatrix}, B_1 = \begin{bmatrix} -2 \\ -1 \end{bmatrix},$$

$$B_2 = \begin{bmatrix} -2 \\ -1 \end{bmatrix}, E = \begin{bmatrix} 0.2 & 0 \\ 0 & 0.1 \end{bmatrix}, B_1^w = \begin{bmatrix} 0.36 \\ 0.28 \end{bmatrix}, B_2^w = \begin{bmatrix} 0.19 \\ 0.06 \end{bmatrix},$$

$$C_1 = \begin{bmatrix} 2.3344 & -0.8029 \\ -0.8029 & 2.3485 \\ 0 & 0 \end{bmatrix}, D_1 = \begin{bmatrix} 0 \\ 0 \\ 1.3460 \end{bmatrix},$$

$$C_2 = \begin{bmatrix} 1.8385 & -1.2728 \\ -1.2728 & 1.6971 \\ 0 & 0 \end{bmatrix}, D_2 = \begin{bmatrix} 0 \\ 0 \\ 1.0540 \end{bmatrix}, H_{11} = \begin{bmatrix} 0.03 & 0 \\ 0 & 0.04 \end{bmatrix},$$

$$H_{12} = \begin{bmatrix} 0.02 & 1 \\ 0 & 0.05 \end{bmatrix}, H_{21} = \begin{bmatrix} 0 \\ 0.02 \end{bmatrix}, H_{22} = \begin{bmatrix} 0.01 \\ 0.02 \end{bmatrix},$$

$$\Pi_1 = \begin{bmatrix} 0.1 & 0.9 \\ 0.2 & 0.8 \end{bmatrix}, \Pi_2 = \begin{bmatrix} 0 & 1 \\ 1 & 0 \end{bmatrix}.$$

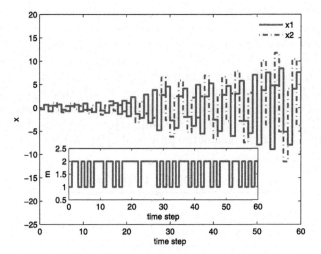

FIGURE 9.6 State trajectories of input-free NMJLS with uncertainties.

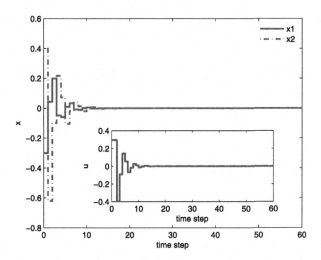

FIGURE 9.7 State trajectories of closed-loop NMJLS with uncertainties.

We choose the indicator functions

$$\alpha_1(k) = \frac{1 - \exp\left(x_1^2\right)}{2}, \alpha_2(k) = \frac{1 + \exp\left(x_1^2\right)}{2}.$$

The state trajectories of the input-free uncertain NMJLS (9.1), and the controlled uncertain NMJLS (9.1) are drawn in Figs. 9.6 and 9.7, respectively. We can see that the input-free NMJLS (9.1) is not stable. However, the trajectory can be regulated to converge to the zero point in the mean square sense by employing the proposed control strategy, which satisfactorily justifies the stochastic stability of the closed-loop NMJLS (9.1).

By solving LMIs (9.35a)–(9.35d) presented in Theorem 9.3, the mode-independent controller gain K and optimized attenuation level γ can be obtained as $K = [-0.1440 \quad 0.6281]$ and $\gamma = 1.6428$, respectively. A simple comparison can also be made by employing the design approach presented in [7], i.e., $K = [-0.2159 \quad 0.5340]$ and $\gamma = 2.3287$, which satisfactorily justifies the advantage of the proposed approach.

9.7 CONCLUSION

In this chapter, we investigate the problems of stochastic stability and H_∞ performance analysis for a class of NMJLSs. By allowing the multistep LF to increase during several sampling time step within the jump mode we obtain a more general stability criteria with better disturbance attenuation capability. Via LMI formulation, we derive existence conditions of the robust H_∞ controller for the underlying systems with time-varying uncertainties in the multistep case. Our main results also have potentials to deal with the model predictive control problems for NMJLSs.

References

[1] L.X. Zhang, T. Yang, P. Shi, M. Liu, Stability and stabilization of a class of discrete-time fuzzy systems with semi-Markov stochastic uncertainties, IEEE Transactions on Systems, Man, and Cybernetics: Systems 46 (12) (2016) 1642–1653.
[2] L.V. Costa, M.D. Fragoso, R.P. Marques, Continuous Time Markovian Jump Linear Systems, Springer, Berlin, 2013.
[3] G.L. Wang, Q.L. Zhang, X.G. Yan, Analysis and Design of Singular Markovian Jump Systems, Springer, Switzerland, 2015.
[4] Z.T. Hou, J.W. Luo, P. Shi, S.K. Nguang, Stochastic stability of Ito differential equations with semi-Markovian jump parameters, IEEE Transactions on Automatic Control 51 (8) (2006) 1383–1387.
[5] J. Huang, Y. Shi, Stochastic stability and robust stabilization of semi-Markov jump linear systems, International Journal of Robust and Nonlinear Control 23 (18) (2013) 2028–2043.
[6] F.B. Li, L.G. Wu, P. Shi, C.C. Lim, State estimation and sliding mode control for semi-Markovian jump systems with mismatched uncertainties, Automatica 51 (2015) 385–393.
[7] S. Aberkane, Stochastic stabilization of a class of nonhomogeneous Markovian jump linear systems, Systems & Control Letters 60 (3) (2011) 156–160.
[8] P. Bolzern, P. Colaneri, G. De Nicolao, Markov jump linear systems with switching transition rates: mean square stability with dwell-time, Automatica 46 (6) (2010) 1081–1088.
[9] J.W. Wen, L. Peng, S.K. Nguang, Finite-time analysis and design for discrete-time switching dynamics Markovian jump linear systems with time-varying delay, IET Control Theory & Applications 17 (8) (2014) 1972–1985.
[10] J.W. Wen, L. Peng, S.K. Nguang, Stochastic finite-time boundedness on switching dynamics Markovian jump linear systems with saturated and stochastic nonlinearities, Information Sciences 334 (2016) 65–82.
[11] L.X. Zhang, H_∞ estimation for discrete-time piecewise homogeneous Markov jump linear systems, Automatica 45 (11) (2009) 2570–2576.

[12] Y.Y. Yin, P. Shi, F. Liu, K.L. Teo, Observer-based H_∞ control on nonhomogeneous Markov jump systems with nonlinear input, International Journal of Robust and Nonlinear Control 24 (13) (2014) 1903–1924.

[13] Y.Y. Yin, P. Shi, F. Liu, K.L. Teo, Filtering for discrete-time nonhomogeneous Markov jump systems with uncertainties, Information Sciences 259 (2014) 118–127.

[14] Y.Y. Yin, P. Shi, F. Liu, K.L. Teo, Robust fault detection for discrete-time stochastic systems with non-homogeneous jump processes, IET Control Theory & Applications 8 (1) (2014) 1–10.

[15] Y.Y. Yin, P. Shi, F. Liu, K.L. Teo, C.C. Lim, Robust filtering for nonlinear nonhomogeneous Markov jump systems by fuzzy approximation approach, IEEE Transactions on Cybernetics 45 (9) (2015) 1706–1716.

[16] F.B. Li, L.G. Wu, P. Shi, Stochastic stability of semi-Markovian jump systems with mode-dependent delays, International Journal of Robust and Nonlinear Control 24 (18) (2014) 3317–3330.

[17] L.X. Zhang, Y.S. Leng, P. Colaneri, Stability and stabilization of discrete-time semi-Markov jump linear systems via semi-Markov kernel approach, IEEE Transactions on Automatic Control 61 (2) (2016) 503–508.

[18] Y.C. Ding, H. Liu, Stability analysis of continuous-time Markovian jump time-delay systems with time-varying transition rates, Journal of the Franklin Institute 353 (11) (2016) 2418–2430.

[19] C.K. Ahn, P. Shi, M.V. Basin, Two-dimensional dissipative control and filtering for Roesser model, IEEE Transactions on Automatic Control 60 (7) (2015) 1745–1759.

[20] C.K. Ahn, L.G. Wu, P. Shi, Stochastic stability analysis for 2-D Roesser systems with multiplicative noise, Automatica 69 (2016) 356–363.

[21] C.K. Ahn, P. Shi, M.V. Basin, Deadbeat dissipative FIR filtering, IEEE Transactions on Circuits and Systems I: Regular Papers 63 (8) (2016) 1210–1221.

[22] A.A. Ahmadi, P.A. Parrilo, Non-monotonic Lyapunov functions for stability of discrete time nonlinear and switched systems, in: Proceedings of 47th IEEE Conference on Decision and Control, 2008, pp. 614–621.

[23] S.F. Derakhshan, A. Fatehi, Non-monotonic Lyapunov functions for stability analysis and stabilization of discrete time Takagi–Sugeno fuzzy systems, International Journal of Innovative Computing, Information & Control (2014) 1567–1586.

[24] S.F. Derakhshan, A. Fatehi, M.G. Sharabiany, Nonmonotonic observer-based fuzzy controller designs for discrete time T-S fuzzy systems via LMI, IEEE Transactions on Cybernetics 44 (12) (2014) 2557–2567.

[25] S.F. Derakhshan, A. Fatehi, Non-monotonic robust H_2 fuzzy observer-based control for discrete time nonlinear systems with parametric uncertainties, International Journal of Systems Science 46 (12) (2015) 2134–2149.

[26] A. Kruszewski, R. Wang, T.M. Guerra, Nonquadratic stabilization conditions for a class of uncertain nonlinear discrete time TS fuzzy models: a new approach, IEEE Transactions on Automatic Control 53 (2) (2008) 606–611.

[27] Y.J. Chen, M. Tanaka, K. Inoue, H. Ohtake, K. Tanaka, T.M. Guerra, A. Kruszewski, H.O. Wang, A nonmonotonically decreasing relaxation approach of Lyapunov functions to guaranteed cost control for discrete fuzzy systems, IET Control Theory & Applications 8 (16) (2014) 1716–1722.

[28] A. Nasiri, S.K. Nguang, A. Swain, D.J. Almakhles, Robust output feedback controller design of discrete-time Takagi–Sugeno fuzzy systems: a non-monotonic Lyapunov approach, IET Control Theory & Applications 10 (5) (2016) 545–553.

[29] D.R. Ding, Z.D. Wang, B. Shen, G.L. Wei, Event-triggered consensus control for discrete-time stochastic multi-agent systems: the input-to-state stability in probability, Automatica 62 (2015) 284–291.

[30] S. Liu, G.L. Wei, Y. Song, Y.R. Liu, Error-constrained reliable tracking control for discrete time-varying systems subject to quantization, Neurocomputing 174 (2016) 897–905.

[31] Q. Li, B. Shen, Y.R. Liu, F.E. Alsaadi, Event-triggered H_∞ state estimation for discrete-time stochastic genetic regulatory networks with Markovian jumping parameters and time-varying delays, Neurocomputing 174 (2016) 912–920.

[32] S.K. Nguang, W. Assawinchaichote, P. Shi, Y. Shi, Robust H_∞ control design for uncertain fuzzy systems with Markovian jumps: an LMI approach, in: American Control Conference, vol. 3, 2005, pp. 1805–1810.

[33] O.L.V. Costa, E.O.A. Filho, E.K. Boukas, R.P. Marques, Constrained quadratic state feedback control of discrete-time Markovian jump linear systems, Automatica 35 (4) (1999) 617–626.

10

Robust H_∞ Filtering of Nonhomogeneous Markovian Jump Delay Systems via N-Step Ahead Lyapunov–Krasovskii Function Approach

10.1 INTRODUCTION

Filtering is one of the most fundamental problems in modern control theory and application fields, such as control system synthesis, state estimation, information fusion, etc. From the signal-processing point of view, useful signals are inevitably contaminated by exogenous noise, which leads to the discrepancy between obtained signals and original signals during the measurement and transmission. It is thus necessary to minimize inaccuracies caused by noise and to estimate the signal close to the original one [1]. Filtering methods are rightly aimed at this purpose. The goal of filtering is estimating signals that are corrupted by noises or unmeasurable or technically difficult to be measured.

With the development of Kalman filtering theory [2] for stochastic systems, numerous extended results of such an optimal filtering have been reported with burgeoning research interest. To name a few, a robust Kalman filter has been designed for continuous-time delay systems with norm bounded uncertainties [3] and discrete-time uncertain systems [4] by fully making use of the mean value and the variance knowledge of Gaussian noise. Nowadays, a new research tendency is studying the Kalman filtering problem under the networked environment. The distributed Kalman

Non-Monotonic Approach to Robust H∞ Control of
Multi-Model Systems
https://doi.org/10.1016/B978-0-12-814868-6.00016-2

filtering with effects of network structures was introduced by a thorough bibliographic review in [5]. Recently, a distributed extended Kalman filter with nonlinear consensus estimate [6] or with event-triggered scheme and stability guarantee [7] has been developed. As a typical application example, formation of autonomous multivehicles was achieved by employing distributed Kalman filter design. Each vehicle aims to estimate its own state, such as position, velocity, and acceleration, by locally available measurements and limited communication with its neighbors [8]. Fading measurements are also noteworthy phenomena induced by the wireless network, where fading rates during the communication are described by random variables with known statistical properties dependent or independent on the system mode. It usually results in unpredictable performance degradation for the filters [9]. The Tobit Kalman filter [10] and modified extended Kalman filter [11] have been further investigated over fading channels with transmission failure or signal fluctuation.

A successful application of Kalman filter in the aerospace and aviation industry has led applications in ordinary industry in the 1970s. However, these attempts have shown that there was a serious mismatch between the underlying assumptions of Kalman filtering and the industrial state estimation problem. In terms of engineering applications, it is quite costly or difficult to get an accurate system model, and engineers are seldom aware of the statistical characteristics of noise that affect the industrial processes. A useful scheme to deal with modeling uncertainty and non-Gaussian noise is H_∞ filtering. The objective is to minimize the l_2 gain of the filtering error system from noise inputs to filtering errors. H_∞ filtering can be effective in the case that noise inputs are energy-bounded signals, without any statistical knowledge of the noise, rather than Gaussian white noise [12]. Moreover, it is robust against uncertainties both in the system modeling and exogenous noise.

The development of H_∞ filtering is reviewed in [13–16] and references therein. It is worth mentioning that H_∞ filtering for a class of special switched systems, i.e., Markovian jump systems (MJSs), attracted much attention of the control community. MJSs can well depict various physical phenomena, such as solar thermal receiver, Samuelson's multiplier-accelerator model, NASA F-8 test aircraft, and networked control systems [17]. Such systems may run with external environmental changes, actuator fault, communication time delay, data packet loss, and so on. These random factors often cause a jumping phenomenon of system structure or parameters. Aiming at the H_∞ filtering for MJSs, a great deal of research work has emerged (see [17–19] and references therein). The essential difference between MJSs and linear systems is the modes jumping character, which is governed by the transition probabilities (TPs). Therefore, the H_∞ filtering for nonhomogeneous Markovian jump systems (NMJSs), that is, MJSs with time-varying TPs, starts to penetrate into the research frontline of the filtering. Filtering for discrete-time uncertain NMJSs [20], robust

H_∞ filtering for continuous-time nonlinear NMJSs with randomly occurring uncertainties [21], and finite-time H_∞ filtering for nonlinear singular NMJSs [22] have been extensively studied. Moreover, the H_∞ filtering for time-delayed systems has also attracted great research interests (see [38–40] and references therein). Many effective approaches, such as the famous Lyapunov–Krasovskii function (LKF) approach and Lyapunov–Razumikhin function approach are developed independently for handling time delay [41]. Specifically, robust filtering performance can be achieved by constructing a proper LKF for time-delay systems.

However, filter design approaches developed for NMJSs in the above-mentioned work are conservative. To reduce the design conservatism, an interesting idea is to construct nonmonotonic Lyapunov function (NLF) [23]. It has been successfully used for T-S fuzzy models to relax the monotonicity requirement of LF and further reduce the conservatism of the stability criteria, i.e., allowing the LF to increase locally during several sampling periods. Two-samples variation, i.e., $V(x_{k+2}) < V(x_k)$ [24–26], and N-sample variation, i.e., $V(x_{k+N}) < V(x_k)$ [27–33], were fully developed for the T-S fuzzy model. Stability analysis and synthesis, robust H_∞ controller design, observer-based fuzzy controller design, and output feedback stabilization have been intensively studied. Some preliminary results for switched systems [34,35] and NMJSs [36] have also been reported.

Based on these observations, we can conclude that the NLF approach has not been fully developed for filtering of time-delayed systems. The bigger challenge is how to deal with noise of future time and delay intervals according to the predictive horizon N. In this study, we aim to investigate the H_∞ filtering problem for a class of nonhomogeneous Markovian jump delay systems (NMJDSs) via an N-step ahead Lyapunov–Krasovskii function (NALKF) approach. The main contributions and novelties of this chapter are summarized as follows:

• The NALKF approach is developed to reduce the conservatism of the filtering design by properly constructing an LKF and allowing the underlying LKF to increase during the period of N sampling time steps ahead of the current time within each jump mode.

• The LMI formulation of sufficient conditions for filtering is obtained by moving the horizon from $k + N - 1$ to $k + 1$ step by step. Due to the predictive horizon, the derivation is not trivial.

• For all possible time-varying TPs and all admissible parameter uncertainties and time delays, the filtering error system is mean-square stable with smaller estimated error and lower dissipative level.

Notations: $\mathbb{R}^n$ and $\mathbb{R}^{n \times m}$ denote the n-dimensional Euclidean space and the set of all $n \times m$ matrices, respectively. $\| \cdot \|_2$ refers to the Euclidean vector norm; $\mathscr{E}\{\cdot \mid \mathscr{F}_k\}$ stands for the conditional mathematical expectation, where $\mathscr{F}_k = \sigma\{(x_0, r_0), \ldots, (x_k, r_k)\}$ is the σ-algebra; $\lambda_{\min}(\Theta)$ represents the minimum eigenvalue of matrix Θ; and $\text{diag}_n(\cdots)$ stands for the block-diagonal matrix with n blocks given by matrices in $(\cdots)$. In symmetric matrices, we use $*$ as an ellipsis for the symmetric terms above or below the diagonal. The identity and zero matrices of appropriate dimensions are denoted by $\mathbf{I}$ and $\mathbf{0}$, respectively; $\mathbb{Z}_{[s_1,s_2]} \triangleq \{l | l \in \mathbb{Z}, s_1 \leq l \leq s_2\}$, where $\mathbb{Z}$ is the set of integers; and $l_{2[0,S]}$ is the space of summable sequences on $\mathbb{Z}_{[0,S]}$, where S may be finite or infinite.

10.2 SYSTEM DESCRIPTION AND PROBLEM FORMULATION

Given the complete probability space $(\Omega, \mathcal{F}, \mathscr{P})$, we consider a discrete-time NMJDS described by the following stochastic difference equation:

$$\begin{cases} x_{k+1} = (A(r_k) + \delta A(r_k)) x_k \\ \qquad + (A_d(r_k) + \delta A_d(r_k)) x_{k-d} + B(r_k)w_k, \\ y_k = C(r_k)x_k + D(r_k)w_k, \\ z_k = G(r_k)x_k + F(r_k)w_k, \\ x_k = \phi_k, k = -d, -d+1, \ldots, 0, \end{cases} \qquad (10.1)$$

where $k \in \mathbb{Z}_{[0,M]}$, $x_k \in \mathbb{R}^{n_x}$ is the state vector, $y_k \in \mathbb{R}^{n_y}$ is the measured output, $z_k \in \mathbb{R}^{n_z}$ is the output to be estimated, $w_k \in \mathbb{R}^{n_w}$ is the exogenous disturbances satisfying $w_k \in l_{2[0,S]}$, the integer $d > 0$ is the time delay, and ϕ_k is a real-valued initial function. When $w_k = 0$, system (10.1) is usually named as a *disturbance-free* NMJDS.

For each possible value of $r_k = i_0$, we denote the matrices associated with the "i_0th jump mode" by $\mathscr{A}_{i_0}$, where $\mathscr{A}$ can be replaced by any symbol. The matrices A_{i_0}, A_{d,i_0}, B_{i_0}, C_{i_0}, D_{i_0}, G_{i_0}, and F_{i_0} are known constant matrices depending on the jump mode r_k. The matrices δA_{i_0} and $\delta A_{d,i_0}$ represent time-varying parameter uncertainties, which are assumed to be norm bounded and can be noted as

$$[\delta A_{i_0} \quad \delta A_{d,i_0}] = E_{i_0} \mathscr{L}_{i_0}(k)[H_{i_0}^a \quad H_{i_0}^d], \qquad (10.2)$$

where E_{i_0}, $H_{i_0}^a$, and $H_{i_0}^d$ are known constant matrices, which characterize the structure of uncertainties. The matrices $\mathscr{L}_{i_0}(k)$ are unknown time-varying matrix functions with Lebesgue-measurable elements satisfying $\mathscr{L}_{i_0}^{\mathrm{T}}(k)\mathscr{L}_{i_0}(k) \leq I$.

The symbol r_k is a discrete-time discrete-state Markov chain taking values in $\mathbb{Z}_{[1,\beta]}$ with transition probabilities

$$\Pr(r_{k+1} = i_1 | r_k = i_0) = \pi_{i_0 i_1}(k), \qquad (10.3)$$

where $\pi_{i_0 i_1}(k) \geq 0$ satisfy $\sum_{i_1=1}^{\beta} \pi_{i_0 i_1}(k) = 1$.

The symbols $\pi_{i_0 i_1}(k) \triangleq \sum_{f_0=1}^{\rho} \mu_{f_0}(k) \pi_{i_0 i_1}^{f_0}$ ($f_0 \in \mathbb{Z}_{[1,\rho]}$) are the entries of the TP matrix $\Pi(k)$ defined as

$$\Pi(k) = \Pi(\mu(k)) \triangleq \sum_{f_0=1}^{\rho} \mu_{f_0}(k) \Pi_{f_0}, \tag{10.4}$$

where $\mu_{f_0}(k) \geq 0$, $\sum_{f_0=1}^{\rho} \mu_{f_0}(k) = 1$, and Π_{f_0} are given TP matrices. All polytopic indices of the indicator function $\mu(k)$ over the time interval $[k, k+N]$ can be boiled down to a set $\mathbb{F}_{[0,N]} \triangleq \{f_0, f_1, \ldots, f_N\}$, where $f_0, f_1, \ldots, f_N \in \mathbb{Z}_{[1,\rho]}$.

Since the N-step ahead scenario will be considered, we further denote a chain of values with respect to the current time jump mode r_k to the future jump mode r_{k+N}. Let $r_k \triangleq i_0$, $r_{k+1} \triangleq i_1, \ldots, r_{k+N} \triangleq i_N$, and then we have a series of value sets for finite-state Markov chains, $\mathbb{I}_{0,l} \triangleq \{i_0, i_1, \cdots, i_l\}$ ($l \in \mathbb{Z}_{[1,N]}$), over the time interval $[k, k+N]$, where $\{i_0, i_1 \ldots, i_N\} \in \mathbb{Z}_{[1,\beta]}$.

In this chapter, we are interested in designing the following mode-dependent full-order filter for system (10.1):

$$\begin{cases} \hat{x}_{k+1} = A_F(r_k)\hat{x}_k + B_F(r_k)y_k, \\ \hat{z}_k = C_F(r_k)\hat{x}_k + D_F(r_k)y_k, \end{cases} \tag{10.5}$$

where $\hat{x}_k \in \mathbb{R}^{n_x}$ and $\hat{z}_k \in \mathbb{R}^{n_z}$ are the estimated state and estimate of the output z_k, respectively. For $r_k = i_0$, the matrices A_{F,i_0}, B_{F,i_0}, C_{F,i_0}, and D_{F,i_0} are filter gains to be determined.

By introducing new vectors $\xi_k = [x_k^{\mathrm{T}} \quad \hat{x}_k^{\mathrm{T}}]^{\mathrm{T}}$ and $e_k = z_k - \hat{z}_k$ the resulting filtering error system of system (10.1) together with filter (10.5) becomes

$$\begin{cases} \xi_{k+1} = \hat{A}_{i_0}\xi_k + \hat{A}_{d,i_0}\xi_{k-d} + \hat{B}_{i_0}w_k, \\ e_k = \hat{C}_{i_0}\xi_k + \hat{D}_{i_0}w_k, \end{cases} \tag{10.6}$$

where

$$\hat{A}_{i_0} = \left[\begin{array}{c:c} \tilde{A}_{i_0} & 0 \\ \hdashline B_{F,i_0}C_{i_0} & A_{F,i_0} \end{array} \right], \quad \hat{A}_{d,i_0} = \left[\begin{array}{c:c} \tilde{A}_{d,i_0} & 0 \\ \hdashline 0 & 0 \end{array} \right],$$

$$\hat{B}_{i_0} = \left[\begin{array}{c} B_{i_0} \\ \hdashline B_{F,i_0}D_{i_0} \end{array} \right], \quad \hat{C}_{i_0} = \left[\begin{array}{c:c} G_{i_0} - D_{F,i_0}C_{i_0} & -C_{F,i_0} \end{array} \right],$$

$$\hat{D}_{i_0} = F_{i_0} - D_{F,i_0}D_{i_0}, \quad \tilde{A}_{i_0} = A_{i_0} + E_{i_0}\mathcal{L}_{i_0}(k)H_{i_0}^a,$$

$$\tilde{A}_{d,i_0} = A_{d,i_0} + E_{i_0}\mathcal{L}_{i_0}(k)H_{i_0}^d. \tag{10.7}$$

To deal with the mean-square stability of the filtering error system (10.6), we give the following preliminaries.

Definition 10.1. The *disturbance-free* NMJDS (10.6) is said to be mean-square stable if

$$\lim_{S \to \infty} \sum_{k=0}^{S} \mathscr{E} \left\{ \|\xi_k\|_2^2 \right\} < \infty \tag{10.8}$$

for any finite initial conditions $\xi_0 \in \mathbb{R}^{2n_x}$ and $r_0 \in \mathbb{Z}_{[1,\beta]}$.

Assumption 10.1. NMJDS (10.1) is mean-square stable for all admissible uncertainties (10.2).

Assumption 10.1 is made based on the fact that NMJDS (10.1) is autonomous without control inputs. Therefore, the mean-square stability of the original system (10.1) to be estimated must be a prerequisite for the mean-square stability of the filtering error system (10.6).

The purpose of this chapter is to design H_∞ filter of the form (10.5) such that the following two conditions are satisfied:

i) The filtering error system (10.6) with $w_k = 0$ is mean-square stable;

ii) Under the zero initial condition, the filtering error e satisfies

$$\sum_{k=0}^{S} \mathscr{E} \left\{ \|e_k\|_2^2 \right\} \leqslant \gamma^2 \sum_{k=0}^{S} \|w_k\|_2^2 \tag{10.9}$$

for any nonzero $w_k \in l_{2[0,S]}$ and a given/optimized l_2 gain bound $\gamma > 0$.

10.3 H_∞ FILTERING PERFORMANCE ANALYSIS

In this section, we first provide a relaxed H_∞ analysis result for the filtering error system (10.6) by employing the NALKF approach, which will be used for filter design in the next section.

Lemma 10.1. *The disturbance-free NMJDS (10.6) is mean-square stable if there exist a set of symmetric matrices* $M_{i_l}^{f_l}$, $P_{i_l}^{f_l} = \left(P_{i_l}^{f_l} \right)^{\mathrm{T}} > 0$ *for all the* $i_l \in \mathbb{Z}_{[1,\beta]}$, $f_l \in \mathbb{Z}_{[1,\rho]}$, $l \in \mathbb{Z}_{[0,N]}$, $q \in \mathbb{Z}_{[1,N-1]}$, *and* $Q = Q^{\mathrm{T}} > 0$ *such that*

$$\Theta_N = \begin{bmatrix} \Theta_{N,11} & \Theta_{N,12} \\ \hline * & \Theta_{N,22} \end{bmatrix} < 0, \tag{10.10a}$$

$$\Theta_{\mathbb{Z}_{[1,N-1]}} = \begin{bmatrix} \Theta_{\mathbb{Z}_{[1,N-1]},11} & \Theta_{\mathbb{Z}_{[1,N-1]},12} \\ \hline * & \Theta_{\mathbb{Z}_{[1,N-1]},22} \end{bmatrix} < 0, \tag{10.10b}$$

$$\Theta_0 = \begin{bmatrix} \Theta_{0,11} & \Theta_{0,12} \\ \hline * & \Theta_{0,22} \end{bmatrix} < 0, \tag{10.10c}$$

where

$$\Theta_{N,11} = -\tilde{M}_{i_0,i_{N-1}}^{f_{N-2},f_{N-1}} + \hat{A}_{i_{N-1}}^{\mathrm{T}} \tilde{P}_{i_0,i_N}^{f_{N-1},f_N} \hat{A}_{i_{N-1}} + Q,$$

$$\Theta_{N,12} = \hat{A}_{i_{N-1}}^{\mathrm{T}} \tilde{P}_{i_0,i_N}^{f_{N-1},f_N} \hat{A}_{d,i_{N-1}},$$

$$\Theta_{N,22} = -Q + \hat{A}_{d,i_{N-1}}^{\mathrm{T}} \tilde{P}_{i_0,i_N}^{f_{N-1},f_N} \hat{A}_{d,i_{N-1}},$$

$$\Theta_{\mathbb{Z}_{[1,N-1]},11} = -\tilde{M}_{i_0,i_{N-q-1}}^{f_{N-q-2},f_{N-q-1}} + \hat{A}_{i_{N-q-1}}^{\mathrm{T}} \tilde{M}_{i_0,i_{N-q}}^{f_{N-q-1},f_{N-q}} \hat{A}_{i_{N-q-1}} + Q,$$

$$\Theta_{\mathbb{Z}_{[1,N-1]},12} = \hat{A}_{i_{N-q-1}}^{\mathrm{T}} \tilde{M}_{i_0,i_{N-q}}^{f_{N-q-1},f_{N-q}} \hat{A}_{d,i_{N-q-1}},$$

$$\Theta_{\mathbb{Z}_{[1,N-1]},22} = -Q + \hat{A}_{d,i_{N-q-1}}^{\mathrm{T}} \tilde{M}_{i_0,i_{N-q}}^{f_{N-q-1},f_{N-q}} \hat{A}_{d,i_{N-q-1}},$$

$$\Theta_{0,11} = -P_{i_0}^{f_0} + \hat{A}_{i_0}^{\mathrm{T}} \tilde{M}_{i_0,i_1}^{f_0,f_1} \hat{A}_{i_0} + Q,$$

$$\Theta_{0,12} = \hat{A}_{i_0}^{\mathrm{T}} \tilde{M}_{i_0,i_1}^{f_0,f_1} \hat{A}_{d,i_0},$$

$$\Theta_{0,22} = -Q + \hat{A}_{d,i_0}^{\mathrm{T}} \tilde{M}_{i_0,i_1}^{f_0,f_1} \hat{A}_{d,i_0},$$

$$\tilde{M}_{i_0,i_{N-1}}^{f_{N-2},f_{N-1}} = \sum_{i_{N-1}=1}^{\beta} \pi_{i_0,i_{N-1}}^{f_{N-2}} M_{i_{N-1}}^{f_{N-1}},$$

$$\tilde{P}_{i_0,i_N}^{f_{N-1},f_N} = \sum_{i_N=1}^{\beta} \pi_{i_0,i_N}^{f_{N-1}} P_{i_N}^{f_N},$$

$$\tilde{M}_{i_0,i_{N-q-1}}^{f_{N-q-2},f_{N-q-1}} = \sum_{i_{N-q-1}=1}^{\beta} \pi_{i_0,i_{N-q-1}}^{f_{N-q-2}} M_{i_{N-q-1}}^{f_{N-q-1}},$$

$$\tilde{M}_{i_0,i_{N-q}}^{f_{N-q-1},f_{N-q}} = \sum_{i_{N-q}=1}^{\beta} \pi_{i_0,i_{N-q}}^{f_{N-q-1}} M_{i_{N-q}}^{f_{N-q}},$$

$$\tilde{M}_{i_0,i_1}^{f_0,f_1} = \sum_{i_1=1}^{\beta} \pi_{i_0,i_1}^{f_0} M_{i_1}^{f_1}.$$

Proof. Consider the following Lyapunov–Krasovskii function at time k:

$$V(\varphi_k) \triangleq V(\mu(k), r_k, \xi_k)$$

$$\triangleq \xi_k^{\mathrm{T}} \left(\sum_{f_0=1}^{\rho} \mu_{f_0}(k) P_{i_0}^{f_0} \right) \xi_k + \sum_{m=k+1-d}^{k} \xi_m^{\mathrm{T}} Q \xi_m. \quad (10.11)$$

Denote

$$\Omega_{0,l}(k) \triangleq \sum_{f_0=1}^{\rho} \mu_{f_0}(k) \sum_{f_1=1}^{\rho} \mu_{f_1}(k+1) \cdots \sum_{f_l=1}^{\rho} \mu_{f_l}(k+l), \quad (10.12)$$

$$\eta_k \triangleq [\xi_k^T \quad \xi_{k-d}^T]^T, \tag{10.13}$$

$$V'(\varphi_k) \triangleq \xi_k^T \left(\sum_{f_0=1}^{\rho} \mu_{f_0}(k) M_{i_0}^{f_0} \right) \xi_k + \sum_{m=k+1-d}^{k} \xi_m^T Q \xi_m. \tag{10.14}$$

By virtue of Definition 10.1, we develop the following steps to prove that

$$\Delta_N V(\varphi_k) = \mathscr{E}\{V(\varphi_{k+N}) \mid \mathscr{F}_k\} - V(\varphi_k) < 0 \tag{10.15}$$

establishes the mean-square stability of disturbance-free NMJDS (10.6).

Subtracting and adding a group of conditional mathematical expectations $\mathscr{E}\{V'(\varphi_{k+N-1}) \mid \mathscr{F}_k\}, \mathscr{E}\{V'(\varphi_{k+N-2}) \mid \mathscr{F}_k\}, \ldots, \mathscr{E}\{V'(\varphi_{k+1}) \mid \mathscr{F}_k\}$ to and from (10.15), we have

$$
\begin{aligned}
&\Delta_N V(\varphi_k) \\
={}& \mathscr{E}\{V(\varphi_{k+N}) \mid \mathscr{F}_k\} - V(\varphi_k) \\
={}& \mathscr{E}\{V(\varphi_{k+N}) \mid \mathscr{F}_k\} - \mathscr{E}\{V'(\varphi_{k+N-1}) \mid \mathscr{F}_k\} \\
&+ \mathscr{E}\{V'(\varphi_{k+N-1}) \mid \mathscr{F}_k\} - \mathscr{E}\{V'(\varphi_{k+N-2}) \mid \mathscr{F}_k\} \\
&+ \cdots + \mathscr{E}\{V'(\varphi_{k+1}) \mid \mathscr{F}_k\} - V(\varphi_k).
\end{aligned}
\tag{10.16}
$$

Step 1: Bearing in mind that

$$\xi_{k+N} = \hat{A}_{i_{N-1}}\xi_{k+N-1} + \hat{A}_{d,i_{N-1}}\xi_{k-d+N-1},$$

we have

$$
\begin{aligned}
&\mathscr{E}\{V(\varphi_{k+N}) \mid \mathscr{F}_k\} - \mathscr{E}\{V'(\varphi_{k+N-1}) \mid \mathscr{F}_k\} \\
={}& \left(\hat{A}_{i_{N-1}}\xi_{k+N-1} + \hat{A}_{d,i_{N-1}}\xi_{k-d+N-1} \right)^T \Omega_{0,N-1}(k) \\
&\times \tilde{P}_{i_0,i_N}^{f_{N-1},f_N} \left(\hat{A}_{i_{N-1}}\xi_{k+N-1} + \hat{A}_{d,i_{N-1}}\xi_{k-d+N-1} \right) \\
&- \xi_{k+N-1}^T \Omega_{0,N-2}(k) \tilde{M}_{i_0,i_{N-1}}^{f_{N-2},f_{N-1}} \xi_{k+N-1} \\
&+ \xi_{k+N-1}^T Q \xi_{k+N-1} - \xi_{k-d+N-1}^T Q \xi_{k-d+N-1} \\
={}& \eta_{k+N-1}^T \Omega_{0,N-1}(k) \Theta_N \eta_{k+N-1}.
\end{aligned}
\tag{10.17}
$$

We can infer from (10.10a) that

$$\mathscr{E}\{V(\varphi_{k+N}) \mid \mathscr{F}_k\} - \mathscr{E}\{V'(\varphi_{k+N-1}) \mid \mathscr{F}_k\} < 0. \tag{10.18}$$

Therefore, we have

$$
\begin{aligned}
\Delta_N V(\varphi_k) <{}& \mathscr{E}\{V'(\varphi_{k+N-1}) \mid \mathscr{F}_k\} - \mathscr{E}\{V'(\varphi_{k+N-2}) \mid \mathscr{F}_k\} \\
&+ \cdots + \mathscr{E}\{V'(\varphi_{k+1}) \mid \mathscr{F}_k\} - V(\varphi_k).
\end{aligned}
\tag{10.19}
$$

Step 2: Bearing in mind that

$$\xi_{k+N-1} = \hat{A}_{i_{N-2}}\xi_{k+N-2} + \hat{A}_{d,i_{N-2}}\xi_{k-d+N-2},$$

we obtain

$$
\begin{aligned}
& \mathscr{E}\left\{V'\left(\varphi_{k+N-1}\right) \mid \mathscr{F}_k\right\} - \mathscr{E}\left\{V'\left(\varphi_{k+N-2}\right) \mid \mathscr{F}_k\right\} \\
=\ & \left(\hat{A}_{i_{N-2}}\xi_{k+N-2} + \hat{A}_{d,i_{N-2}}\xi_{k-d+N-2}\right)^{\mathrm{T}}\Omega_{0,N-2}(k) \\
& \times \tilde{P}_{i_0,i_{N-1}}^{f_{N-2},f_{N-1}}\left(\hat{A}_{i_{N-2}}\xi_{k+N-2} + \hat{A}_{d,i_{N-2}}\xi_{k-d+N-2}\right) \\
& -\xi_{k+N-2}^{\mathrm{T}}\Omega_{0,N-3}(k)\tilde{M}_{i_0,i_{N-2}}^{f_{N-3},f_{N-2}}\xi_{k+N-2} \\
& +\xi_{k+N-2}^{\mathrm{T}}Q\xi_{k+N-2} - \xi_{k-d+N-2}^{\mathrm{T}}Q\xi_{k-d+N-2} \\
=\ & \eta_{k+N-2}^{\mathrm{T}}\Omega_{0,N-2}(k)\Theta_1\eta_{k+N-2}.
\end{aligned}
\tag{10.20}
$$

If (10.10b) holds for $q = 1$, then it results in

$$\mathscr{E}\left\{V'\left(\varphi_{k+N-1}\right) \mid \mathscr{F}_k\right\} - \mathscr{E}\left\{V'\left(\varphi_{k+N-2}\right) \mid \mathscr{F}_k\right\} < 0, \tag{10.21}$$

which further implies

$$
\begin{aligned}
& \Delta_N V\left(\varphi_k\right) \\
< & \mathscr{E}\left\{V'\left(\varphi_{k+N-2}\right) \mid \mathscr{F}_k\right\} - \mathscr{E}\left\{V'\left(\varphi_{k+N-3}\right) \mid \mathscr{F}_k\right\} \\
& +\cdots + \mathscr{E}\left\{V'\left(\varphi_{k+1}\right) \mid \mathscr{F}_k\right\} - V\left(\varphi_k\right).
\end{aligned}
\tag{10.22}
$$

Repeating the procedure $N - 2$ times yields

$$\Delta_N V\left(\varphi_k\right) < \mathscr{E}\left\{V'\left(\varphi_{k+1}\right) \mid \mathscr{F}_k\right\} - V\left(\varphi_k\right). \tag{10.23}$$

Step 3: Bearing in mind that

$$\xi_{k+1} = \hat{A}_{i_0}\xi_k + \hat{A}_{d,i_0}\xi_{k-d},$$

we have

$$
\begin{aligned}
& \mathscr{E}\left\{V'\left(\varphi_{k+1}\right) \mid \mathscr{F}_0\right\} - V\left(\varphi_k\right) \\
=\ & \left(\hat{A}_{i_0}\xi_k + \hat{A}_{d,i_0}\xi_{k-d}\right)^{\mathrm{T}}\Omega_{0,0}(k)\tilde{M}_{i_0,i_1}^{f_0,f_1} \\
& \times \left(\hat{A}_{i_0}\xi_k + \hat{A}_{d,i_0}\xi_{k-d}\right) \\
& -\xi_k^{\mathrm{T}}P_{i_0}^{f_0}\xi_k + \xi_k^{\mathrm{T}}Q\xi_k - \xi_{k-d}^{\mathrm{T}}Q\xi_{k-d} \\
=\ & \eta_k^{\mathrm{T}}\Omega_{0,0}(k)\Theta_0\eta_k.
\end{aligned}
\tag{10.24}
$$

We can infer from (10.10c) that

$$\mathcal{E}\left\{V'\left(\varphi_{k+1}\right) \mid \mathscr{F}_0\right\} - V\left(\varphi_k\right) < 0, \tag{10.25}$$

which implies that

$$\eta_k^T \Omega_{0,0}(k)\Theta_0 \eta_k < 0. \tag{10.26}$$

Step 4: By observing (10.26) we obtain

$$\Delta_N V\left(\varphi_k\right) < \eta_k^T \Omega_{0,0}(k)\Theta_0 \eta_k \leq -\alpha_{i_0}\eta_k^T \eta_k \leq -\alpha \eta_k^T \eta_k, \tag{10.27}$$

where $\alpha_{i_0} = \min_k \left\{\lambda_{\min}\left(\Omega_{0,0}(k)\Theta_0\right)\right\}$ and $\alpha = \min_{i_0 \in \mathbb{Z}_{[1,\beta]}} \alpha_{i_0}$.

Letting $L(\varphi(k)) = \sum_{l=0}^{N-1} V(\varphi(k+l))$, we have

$$V(\varphi(k+N)) - V(\varphi(k)) < 0 \Leftrightarrow L(\varphi(k+1)) - L(\varphi(k)) < 0. \tag{10.28}$$

Therefore, (10.27) and (10.28) yield

$$\mathcal{E}\left\{L\left(\varphi_{k+1}\right) \mid \mathscr{F}_0\right\} - \mathcal{E}\left\{L\left(\varphi_k\right) \mid \mathscr{F}_0\right\} \leq -\alpha \eta_k^T \eta_k,$$

$$\Rightarrow \quad \mathcal{E}\left\{\sum_{k=0}^{S} \Delta_N V\left(\varphi_k\right) \mid \mathscr{F}_0\right\} = \mathcal{E}\left\{L\left(\varphi_{S+1}\right) \mid \mathscr{F}_0\right\} - L\left(\varphi_0\right)$$

$$\leq -\alpha \mathcal{E}\left\{\sum_{k=0}^{S} \|\eta_k\|_2^2 \mid \mathscr{F}_0\right\}$$

$$\Rightarrow \quad \mathcal{E}\left\{\sum_{k=0}^{S} \|\eta_k\|_2^2 \mid \mathscr{F}_0\right\}$$

$$\leq \alpha^{-1}\left\{L\left(\varphi_0\right) - \mathcal{E}\left\{L\left(\varphi_{S+1}\right) \mid \mathscr{F}_0\right\}\right\}$$

$$\leq \alpha^{-1} L\left(\varphi_0\right)$$

$$\Rightarrow \quad \lim_{S \to \infty} \mathcal{E}\left\{\sum_{k=0}^{S} \|\eta_k\|_2^2 \mid \mathscr{F}_0\right\} \leq \alpha^{-1} L\left(\varphi_0\right) < \infty.$$

This implies, by Definition 10.1, that the disturbance-free NMJDS (10.6) is mean-square stable. The proof is completed. $\qquad\square$

Remark 10.1. In comparison with the multistep LF approach developed in [30], the proposed NALKF approach uses the N-step ahead TPs $\sum_{i_N=1}^{\beta} \pi_{i_0,i_N}^{i_{N-1}}$ to replace the multiple multiplication $\sum_{i_0=1}^{s} \pi_{i_0 i_1}^{(l_0)} \sum_{i_1=1}^{s} \pi_{i_1 i_2}^{(l_1)} \cdots \sum_{i_{N-1}=1}^{s} \pi_{i_{N-1} i_N}^{(l_{N-1})}$. It greatly reduces the number of decision variables, especially for the time-delayed scenario.

The following lemma provides sufficient conditions under which the filtering error system, NMJDS (10.6), is mean-square stable and the filtering error se_k satisfy the dissipative requirement (10.9).

Lemma 10.2. *The NMJDS (10.6) possesses the γ-disturbance attenuation level in (10.9) for all $w_k \in l_{2[0,S]}$, $w_k \neq 0$, if there exist a set of symmetric matrices $M_{i_l}^{f_l}$, $P_{i_l}^{f_l} = \left(P_{i_l}^{f_l}\right)^{\mathrm{T}} > 0$ for all $i_l \in \mathbb{Z}_{[1,\beta]}$, $f_l \in \mathbb{Z}_{[1,\rho]}$, $l \in \mathbb{Z}_{[0,N]}$, $q \in \mathbb{Z}_{[1,N-1]}$, and $Q = Q^{\mathrm{T}} > 0$ such that*

$$\Lambda_N = \begin{bmatrix} \Lambda_{N,11} & \Lambda_{N,12} & \Lambda_{N,13} \\ * & \Lambda_{N,22} & \Lambda_{N,23} \\ * & * & \Lambda_{N,33} \end{bmatrix} < 0, \tag{10.29a}$$

$$\Lambda_{\mathbb{Z}_{[1,N-1]}} = \begin{bmatrix} \Lambda_{\mathbb{Z}_{[1,N-1]},11} & \Lambda_{\mathbb{Z}_{[1,N-1]},12} & \Lambda_{\mathbb{Z}_{[1,N-1]},13} \\ * & \Lambda_{\mathbb{Z}_{[1,N-1]},22} & \Lambda_{\mathbb{Z}_{[1,N-1]},23} \\ * & * & \Lambda_{\mathbb{Z}_{[1,N-1]},33} \end{bmatrix} < 0, \tag{10.29b}$$

$$\Lambda_0 = \begin{bmatrix} \Lambda_{0,11} & \Lambda_{0,12} & \Lambda_{0,13} \\ * & \Lambda_{0,22} & \Lambda_{0,23} \\ * & * & \Lambda_{0,33} \end{bmatrix} < 0, \tag{10.29c}$$

where

$$\Lambda_{N,11} = -\tilde{M}_{i_0,i_{N-1}}^{f_{N-2},f_{N-1}} + \hat{A}_{i_{N-1}}^{\mathrm{T}} \tilde{P}_{i_0,i_N}^{f_{N-1},f_N} \hat{A}_{i_{N-1}} + Q + \hat{C}_{i_{N-1}}^{\mathrm{T}} \hat{C}_{i_{N-1}},$$

$$\Lambda_{N,12} = \hat{A}_{i_{N-1}}^{\mathrm{T}} \tilde{P}_{i_0,i_N}^{f_{N-1},f_N} \hat{A}_{d,i_{N-1}},$$

$$\Lambda_{N,13} = \hat{A}_{i_{N-1}}^{\mathrm{T}} \tilde{P}_{i_0,i_N}^{f_{N-1},f_N} \hat{B}_{i_{N-1}} + \hat{C}_{i_{N-1}}^{\mathrm{T}} \hat{D}_{i_{N-1}},$$

$$\Lambda_{N,22} = -Q + \hat{A}_{d,i_{N-1}}^{\mathrm{T}} \tilde{P}_{i_0,i_N}^{f_{N-1},f_N} \hat{A}_{d,i_{N-1}},$$

$$\Lambda_{N,23} = \hat{A}_{d,i_{N-1}}^{\mathrm{T}} \tilde{P}_{i_0,i_N}^{f_{N-1},f_N} \hat{B}_{i_{N-1}},$$

$$\Lambda_{N,33} = -\gamma^2 I + \hat{B}_{i_{N-1}}^{\mathrm{T}} \tilde{P}_{i_0,i_N}^{f_{N-1},f_N} \hat{B}_{i_{N-1}} + \hat{D}_{i_{N-1}}^{\mathrm{T}} \hat{D}_{i_{N-1}},$$

$$\Lambda_{\mathbb{Z}_{[1,N-1]},11} = -\tilde{M}_{i_0,i_{N-q-1}}^{f_{N-q-2},f_{N-q-1}} + \hat{A}_{i_{N-q-1}}^{\mathrm{T}} \tilde{M}_{i_0,i_{N-q}}^{f_{N-q-1},f_{N-q}} \hat{A}_{i_{N-q-1}} + Q + \hat{C}_{i_{N-q-1}}^{\mathrm{T}} \hat{C}_{i_{N-q-1}},$$

$$\Lambda_{\mathbb{Z}_{[1,N-1]},12} = \hat{A}_{i_{N-q-1}}^{\mathrm{T}} \tilde{M}_{i_0,i_{N-q}}^{f_{N-q-1},f_{N-q}} \hat{A}_{d,i_{N-q-1}},$$

$$\Lambda_{\mathbb{Z}_{[1,N-1]},13} = \hat{A}_{i_{N-q-1}}^{\mathrm{T}} \tilde{M}_{i_0,i_{N-q}}^{f_{N-q-1},f_{N-q}} \hat{B}_{i_{N-q-1}} + \hat{C}_{i_{N-q-1}}^{\mathrm{T}} \hat{D}_{i_{N-q-1}},$$

$$\Lambda_{\mathbb{Z}_{[1,N-1]},22} = -Q + \hat{A}_{d,i_{N-q-1}}^{\mathrm{T}} \tilde{M}_{i_0,i_{N-q}}^{f_{N-q-1},f_{N-q}} \hat{A}_{d,i_{N-q-1}},$$

$$\Lambda_{\mathbb{Z}_{[1,N-1]},23} = \hat{A}_{d,i_{N-q-1}}^{\mathrm{T}} \tilde{M}_{i_0,i_{N-q}}^{f_{N-q-1},f_{N-q}} \hat{B}_{i_{N-q-1}},$$

$$\Lambda_{\mathbb{Z}_{[1,N-1]},33} = -\gamma^2 \mathbf{I} + \hat{B}_{i_{N-q-1}}^{\mathrm{T}} \tilde{M}_{i_0,i_{N-q}}^{f_{N-q-1},f_{N-q}} \hat{B}_{i_{N-q-1}}$$
$$+ \hat{D}_{i_{N-q-1}}^{\mathrm{T}} \hat{D}_{i_{N-q-1}},$$

$$\Lambda_{0,11} = -P_{i_0}^{f_0} + \hat{A}_{i_0}^{\mathrm{T}} \tilde{M}_{i_0,i_1}^{f_0,f_1} \hat{A}_{i_0} + Q + \hat{C}_{i_0}^{\mathrm{T}} \hat{C}_{i_0},$$

$$\Lambda_{0,12} = \hat{A}_{i_0}^{\mathrm{T}} \tilde{M}_{i_0,i_1}^{f_0,f_1} \hat{A}_{d,i_0},$$

$$\Lambda_{0,13} = \hat{A}_{i_0}^{\mathrm{T}} \tilde{M}_{i_0,i_1}^{f_0,f_1} \hat{B}_{i_0} + \hat{C}_{i_0}^{\mathrm{T}} \hat{D}_{i_0},$$

$$\Lambda_{0,22} = -Q + \hat{A}_{d,i_0}^{\mathrm{T}} \tilde{M}_{i_0,i_1}^{f_0,f_1} \hat{A}_{d,i_0},$$

$$\Lambda_{0,23} = \hat{A}_{d,i_0}^{\mathrm{T}} \tilde{M}_{i_0,i_1}^{f_0,f_1} \hat{B}_{i_0},$$

$$\Lambda_{0,33} = -\gamma^2 \mathbf{I} + \hat{B}_{i_0}^{\mathrm{T}} \tilde{M}_{i_0,i_1}^{f_0,f_1} \hat{B}_{i_0} + \hat{D}_{i_0}^{\mathrm{T}} \hat{D}_{i_0},$$

and $\tilde{M}_{i_0,i_{N-1}}^{f_{N-2},f_{N-1}}$, $\tilde{P}_{i_0,i_N}^{f_{N-1},f_N}$, $\tilde{M}_{i_0,i_{N-q-1}}^{f_{N-q-2},f_{N-q-1}}$, $\tilde{M}_{i_0,i_{N-q}}^{f_{N-q-1},f_{N-q}}$, $\tilde{M}_{i_0,i_1}^{f_0,f_1}$ are as in Lemma 10.1.

Proof. Consider the Lyapunov–Krasovskii function $V(\varphi_k)$ and the auxiliary function $V'(\varphi_k)$ noted in (10.11) and (10.14), respectively. It is clear that inequalities (10.29a)–(10.29c) incorporate sufficient conditions presented in Lemma 10.1, and therefore Lemma 10.2 implies the mean-square stability of the NMJDS (10.6). Next, we direct our attention to the H_∞ performance analysis via the following steps.

Define new vectors $\zeta_k \triangleq [\xi_k^{\mathrm{T}} \;\; \xi_{k-d}^{\mathrm{T}} \;\; w_k^{\mathrm{T}}]^{\mathrm{T}}$ and $\vartheta(k) \triangleq e_k^{\mathrm{T}} e_k - \gamma^2 w_k^{\mathrm{T}} w_k$. We will prove that

$$\Delta_N V(\varphi_k) + \sum_{l=0}^{N-1} \vartheta_{k+l} < 0 \tag{10.30}$$

guarantees H_∞ performance of the NMJDS (10.6).

Adding and subtracting a group of mathematical items

$$-\mathscr{E}\left\{V'(\varphi_{k+N-1}) \mid \mathscr{F}_k\right\} + \vartheta_{k+N-1},$$
$$-\mathscr{E}\left\{V'(\varphi_{k+N-2}) \mid \mathscr{F}_k\right\} + \vartheta_{k+N-2} - \cdots - \mathscr{E}\left\{V'(\varphi_{k+1}) \mid \mathscr{F}_k\right\} + \vartheta_{k+1}$$

to and from $\Delta_N V(\varphi_k)$, we have

$$\Delta_N V(\varphi_k)$$
$$= \mathscr{E}\left\{V(\varphi_{k+N}) \mid \mathscr{F}_k\right\} - \mathscr{E}\left\{V'(\varphi_{k+N-1}) \mid \mathscr{F}_k\right\} + \vartheta_{k+N-1}$$

$$+\mathscr{E}\left\{V'\left(\varphi_{k+N-1}\right)\mid\mathscr{F}_k\right\}-\mathscr{E}\left\{V'\left(\varphi_{k+N-2}\right)\mid\mathscr{F}_k\right\}+\vartheta_{k+N-2}$$

$$+\cdots+\mathscr{E}\left\{V'\left(\varphi_{k+1}\right)\mid\mathscr{F}_k\right\}-V\left(\varphi_k\right)+\vartheta_k-\sum_{l=0}^{N-1}\vartheta_{k+l}. \quad (10.31)$$

First, following the lines of the proof of Lemma 10.1, we have that (10.29a) implies that

$$\mathscr{E}\left\{V\left(\varphi_{k+N}\right)\mid\mathscr{F}_k\right\}-\mathscr{E}\left\{V'\left(\varphi_{k+N-1}\right)\mid\mathscr{F}_k\right\}+\vartheta_{k+N-1}$$

$$=\quad \zeta_{k+N-1}^{\mathrm{T}}\Omega_{0,N-1}(k)\Lambda_N\zeta_{k+N-1}<0, \quad (10.32)$$

which further yields

$$\Delta_N V\left(\varphi_k\right)$$

$$<\quad \mathscr{E}\left\{V'\left(\varphi_{k+N-1}\right)\mid\mathscr{F}_k\right\}-\mathscr{E}\left\{V'\left(\varphi_{k+N-2}\right)\mid\mathscr{F}_k\right\}+\vartheta_{k+N-2}$$

$$+\cdots+\mathscr{E}\left\{V'\left(\varphi_{k+1}\right)\mid\mathscr{F}_k\right\}-V\left(\varphi_k\right)+\vartheta_k-\sum_{l=0}^{N-1}\vartheta_{k+l}. \quad (10.33)$$

Then we can conclude form (10.29b) that

$$\mathscr{E}\left\{V\left(\varphi_{k+N-q}\right)\mid\mathscr{F}_k\right\}-\mathscr{E}\left\{V'\left(\varphi_{k+N-q-1}\right)\mid\mathscr{F}_k\right\}+\vartheta_{k+N-q-1}$$

$$=\quad \zeta_{k+N-q-1}^{\mathrm{T}}\Omega_{0,N-q-1}(k)\Lambda_{\mathbb{Z}_{[1,N-1]}}\zeta_{k+N-q-1}<0,\left(\forall q\in\mathbb{Z}_{[1,N-1]}\right),$$

$$(10.34)$$

which implies

$$\Delta_N V\left(\varphi_k\right)\quad<\quad \mathscr{E}\left\{V'\left(\varphi_{k+1}\right)\mid\mathscr{F}_k\right\}-V\left(\varphi_k\right)+\vartheta_k-\sum_{l=0}^{N-1}\vartheta_{k+l}. \quad (10.35)$$

Finally, from (10.29c) we obtain

$$\mathscr{E}\left\{V'\left(\varphi_{k+1}\right)\mid\mathscr{F}_0\right\}-V\left(\varphi_k\right)+\vartheta_k$$

$$=\quad \zeta_k^{\mathrm{T}}\Omega_{0,0}(k)\Lambda_0\zeta_k<0, \quad (10.36)$$

which directly yields (10.30) by combining (10.31), (10.33), (10.35), and (10.36).

In the following, we assume the zero initial conditions $x_{-N}=x_{-N+1}=\cdots=x_0=0$. Define

$$J_S\quad\triangleq\quad\mathscr{E}\left\{\sum_{k=0}^{S}\left(z_k^{\mathrm{T}}z_k-\gamma^2 w_k^{\mathrm{T}}w_k\right)\right\}. \quad (10.37)$$

Inequality (10.30) results in

$$\sum_{k=-N}^{S} \left\{ \Delta_N V(\varphi_k) + \sum_{l=0}^{N-1} \vartheta_{k+l} \right\} < 0$$

$$\Rightarrow \quad \mathscr{E}\left\{ V(\varphi_{S+N}) \mid \mathscr{F}_S \right\} + \mathscr{E}\left\{ V(\varphi_{S+N-1}) \mid \mathscr{F}_{S-1} \right\}$$

$$\underbrace{+ \cdots + \mathscr{E}\left\{ V(\varphi_{S+1}) \mid \mathscr{F}_{-N+S+1} \right\}}_{N}$$

$$- \mathscr{E}\left\{ V(\varphi_{-N}) \mid \mathscr{F}_{-N} \right\} - \mathscr{E}\left\{ V(\varphi_{-N+1}) \mid \mathscr{F}_{-N+1} \right\}$$

$$\underbrace{- \cdots - \mathscr{E}\left\{ V(\varphi_{-1}) \mid \mathscr{F}_{-1} \right\}}_{N}$$

$$+ \sum_{k=-N}^{S} (\vartheta_k + \vartheta_{k+1} + \cdots + \vartheta_{k+N-1}) < 0. \qquad (10.38)$$

As $S \to \infty$, all the negative items tend to zero, and we obtain

$$N V(\varphi_\infty) + N J_\infty < 0,$$

which yields $J_\infty < 0$. Therefore, the dissipative inequality (10.8) holds for $S > 0$. This completes the proof. $\qquad \square$

10.4 ROBUST H_∞ FILTER DESIGN

According to H_∞ performance analysis for the NMJDS (10.6), we are now in the position to provide numerical testable conditions to the robust H_∞ filter design. Before giving the main results, we recall the following well-known lemma.

Lemma 10.3. *[37] Let the matrices* $\Psi = \Psi^T$, E, *and* H *be real matrices of appropriate dimensions, with* $\mathscr{L}(k)$ *satisfying* $\mathscr{L}(k)^T \mathscr{L}(k) \leq \mathbf{I}$. *Then*

$$\Psi + E\mathscr{L}(k)H + H^T \mathscr{L}(k)^T E^T < 0 \qquad (10.39)$$

if and only if there exists a scalar $\varepsilon > 0$ *such that*

$$\Psi + \varepsilon^{-1} E E^T + \varepsilon H^T H < 0 \qquad (10.40)$$

or, equivalently,

$$\begin{bmatrix} \Psi & E & \varepsilon H^T \\ * & -\varepsilon\mathbf{I} & 0 \\ * & * & -\varepsilon\mathbf{I} \end{bmatrix} < 0. \qquad (10.41)$$

In the following theorem, the addressed filter design problem is solvable for all possible time-varying transition probabilities and all admissible parameter uncertainties and time delays if a set of LMIs is feasible.

Theorem 10.1. *The NMJDS (10.6) is mean-square stable, and the dissipative constraint (10.8) with an optimized attenuation level γ is achieved for all nonzero w_k if there exist a set of symmetric matrices $M_{i_l,1}^{f_l}$, $M_{i_l,2}^{f_l}$, $M_{i_l,3}^{f_l}$, X_{i_l}, Y_{i_l}, Z_{i_l}, W_{i_l}, V_{i_l}, $P_{i_l,1}^{f_l} = \left(P_{i_l,1}^{f_l}\right)^{\mathrm{T}} > 0$, $P_{i_l,2}^{f_l} = \left(P_{i_l,2}^{f_l}\right)^{\mathrm{T}} > 0$, $P_{i_l,3}^{f_l} = \left(P_{i_l,3}^{f_l}\right)^{\mathrm{T}} > 0$, $Q_1 = Q_1^{\mathrm{T}} > 0$, $Q_2 = Q_2^{\mathrm{T}} > 0$, $Q_3 = Q_3^{\mathrm{T}} > 0$ for all $i_l \in \mathbb{Z}_{[1,\beta]}$, $f_l \in \mathbb{Z}_{[1,\rho]}$, $l \in \mathbb{Z}_{[0,N]}$, $q \in \mathbb{Z}_{[1,N-1]}$, matrices C_{F,i_l} and D_{F,i_l}, and a scalar ε such that*

$$\Xi_N = \begin{bmatrix} \Xi_{N,11} & 0 & \Xi_{N,13} & \Xi_{N,14} & \Xi_{N,15} \\ * & \Xi_{N,22} & 0 & \Xi_{N,24} & \Xi_{N,25} \\ * & * & \Xi_{N,33} & \Xi_{N,34} & 0 \\ * & * & * & \Xi_{N,44} & \Xi_{N,45} \\ * & * & * & * & \Xi_{N,55} \end{bmatrix} < 0, \qquad (10.42\text{a})$$

$$\Xi_{\mathbb{Z}_{[1,N-1]}} = \begin{bmatrix} \Xi_{\mathbb{Z}_{[1,N-1]},11} & 0 & \Xi_{\mathbb{Z}_{[1,N-1]},13} & \Xi_{\mathbb{Z}_{[1,N-1]},14} & \Xi_{\mathbb{Z}_{[1,N-1]},15} \\ * & \Xi_{\mathbb{Z}_{[1,N-1]},22} & 0 & \Xi_{\mathbb{Z}_{[1,N-1]},24} & \Xi_{\mathbb{Z}_{[1,N-1]},25} \\ * & * & \Xi_{\mathbb{Z}_{[1,N-1]},33} & \Xi_{\mathbb{Z}_{[1,N-1]},34} & 0 \\ * & * & * & \Xi_{\mathbb{Z}_{[1,N-1]},44} & \Xi_{\mathbb{Z}_{[1,N-1]},45} \\ * & * & * & * & \Xi_{\mathbb{Z}_{[1,N-1]},55} \end{bmatrix} < 0, $$

$$(10.42\text{b})$$

$$\Xi_0 = \begin{bmatrix} \Xi_{0,11} & 0 & \Xi_{0,13} & \Xi_{0,14} & \Xi_{0,15} \\ * & \Xi_{0,22} & 0 & \Xi_{0,24} & \Xi_{0,25} \\ * & * & \Xi_{0,33} & \Xi_{0,34} & 0 \\ * & * & * & \Xi_{0,44} & \Xi_{0,45} \\ * & * & * & * & \Xi_{0,55} \end{bmatrix} < 0, \qquad (10.42\text{c})$$

where

$$\Xi_{N,11} = \begin{bmatrix} -\tilde{M}_{i_0,i_{N-1},1}^{f_{N-2},f_{N-1}} + Q_1 & -\tilde{M}_{i_0,i_{N-1},2}^{f_{N-2},f_{N-1}} + Q_2 \\ * & -\tilde{M}_{i_0,i_{N-1},3}^{f_{N-2},f_{N-1}} + Q_3 \end{bmatrix},$$

$$\Xi_{N,13} = \begin{bmatrix} 0 & G_{i_{N-1}}^{\mathrm{T}} - C_{i_{N-1}}^{\mathrm{T}} D_{F,i_{N-1}}^{\mathrm{T}} \\ 0 & -C_{F,i_{N-1}}^{\mathrm{T}} \end{bmatrix},$$

$$\Xi_{N,14} = \begin{bmatrix} \left(X_{i_{N-1}} A_{i_{N-1}}\right)^{\mathrm{T}} & \left(Z_{i_{N-1}} A_{i_{N-1}}\right)^{\mathrm{T}} \\ + \left(W_{i_{N-1}} C_{i_{N-1}}\right)^{\mathrm{T}} & + \left(W_{i_{N-1}} C_{i_{N-1}}\right)^{\mathrm{T}} \\ V_{i_{N-1}}^{\mathrm{T}} & V_{i_{N-1}}^{\mathrm{T}} \end{bmatrix},$$

$$\Xi_{N,15} = \begin{bmatrix} 0 & 0 & 0 & (H^a_{i_{N-1}})^{\mathrm{T}} & 0 & 0 \\ 0 & 0 & 0 & 0 & 0 & 0 \end{bmatrix}, \quad \Xi_{N,22} = \begin{bmatrix} -Q_1 & -Q_2 \\ * & -Q_3 \end{bmatrix},$$

$$\Xi_{N,24} = \left[\begin{array}{c|c} (X_{i_{N-1}} A_{d,i_{N-1}})^{\mathrm{T}} & (Z_{i_{N-1}} A_{d,i_{N-1}})^{\mathrm{T}} \\ \hline 0 & 0 \end{array} \right],$$

$$\Xi_{N,25} = \begin{bmatrix} 0 & 0 & 0 & 0 & 0 & (H^d_{i_{N-1}})^{\mathrm{T}} \\ 0 & 0 & 0 & 0 & 0 & 0 \end{bmatrix},$$

$$\Xi_{N,33} = \left[\begin{array}{c|c} -\gamma^2 \mathbf{I} & F^{\mathrm{T}}_{i_{N-1}} - D^{\mathrm{T}}_{i_{N-1}} D^{\mathrm{T}}_{F,i_{N-1}} \\ \hline * & -\mathbf{I} \end{array} \right],$$

$$\Xi_{N,34} = \left[\begin{array}{c|c} (X_{i_{N-1}} B_{i_{N-1}})^{\mathrm{T}} & (Z_{i_{N-1}} B_{i_{N-1}})^{\mathrm{T}} \\ + (W_{i_{N-1}} D_{i_{N-1}})^{\mathrm{T}} & + (W_{i_{N-1}} D_{i_{N-1}})^{\mathrm{T}} \\ \hline 0 & 0 \end{array} \right],$$

$$\Xi_{N,44} = \left[\begin{array}{c|c} \begin{array}{c} -X_{i_{N-1}} - X^{\mathrm{T}}_{i_{N-1}} \\ + \tilde{P}^{f_{N-1},f_N}_{i_0,i_N,1} \end{array} & \begin{array}{c} -Y_{i_{N-1}} - Z^{\mathrm{T}}_{i_{N-1}} \\ + \tilde{P}^{f_{N-1},f_N}_{i_0,i_N,2} \end{array} \\ \hline * & \begin{array}{c} -Y_{i_{N-1}} - Y^{\mathrm{T}}_{i_{N-1}} \\ + \tilde{P}^{f_{N-1},f_N}_{i_0,i_N,3} \end{array} \end{array} \right],$$

$$\Xi_{N,45} = \begin{bmatrix} X_{i_{N-1}} E_{i_{N-1}} & 0 & X_{i_{N-1}} E_{i_{N-1}} & 0 & 0 & 0 \\ Z_{i_{N-1}} E_{i_{N-1}} & 0 & Z_{i_{N-1}} E_{i_{N-1}} & 0 & 0 & 0 \end{bmatrix},$$

$$\Xi_{N,55} = \mathrm{diag}_6(-\varepsilon \mathbf{I}),$$

$$\Xi_{\mathbb{Z}_{[1,N-1]},11} = \left[\begin{array}{c|c} -\tilde{M}^{f_{N-q-2},f_{N-q-1}}_{i_0,i_{N-q-1},1} + Q_1 & -\tilde{M}^{f_{N-q-2},f_{N-q-1}}_{i_0,i_{N-q-1},2} + Q_2 \\ \hline * & -\tilde{M}^{f_{N-q-2},f_{N-q-1}}_{i_0,i_{N-q-1},3} + Q_3 \end{array} \right],$$

$$\Xi_{\mathbb{Z}_{[1,N-1]},13} = \left[\begin{array}{c|c} 0 & G^{\mathrm{T}}_{i_{N-q-1}} - C^{\mathrm{T}}_{i_{N-q-1}} D^{\mathrm{T}}_{F,i_{N-q-1}} \\ \hline 0 & -C^{\mathrm{T}}_{F,i_{N-q-1}} \end{array} \right],$$

$$\Xi_{\mathbb{Z}_{[1,N-1]},14} = \left[\begin{array}{c|c} \begin{array}{c} (X_{i_{N-q-1}} A_{i_{N-q-1}} \\ + W_{i_{N-q-1}} C_{i_{N-q-1}})^{\mathrm{T}} \end{array} & \begin{array}{c} (Z_{i_{N-q-1}} A_{i_{N-q-1}} \\ + W_{i_{N-q-1}} C_{i_{N-q-1}})^{\mathrm{T}} \end{array} \\ \hline V^{\mathrm{T}}_{i_{N-q-1}} & V^{\mathrm{T}}_{i_{N-q-1}} \end{array} \right],$$

$$\Xi_{\mathbb{Z}_{[1,N-1]},15} = \begin{bmatrix} 0 & 0 & 0 & (H^a_{i_{N-q-1}})^{\mathrm{T}} & 0 & 0 \\ 0 & 0 & 0 & 0 & 0 & 0 \end{bmatrix},$$

$$\Xi_{\mathbb{Z}_{[1,N-1]},22} = \begin{bmatrix} -Q_1 & -Q_2 \\ * & -Q_3 \end{bmatrix},$$

$$\Xi_{\mathbb{Z}_{[1,N-1]},24} = \left[\begin{array}{c:c} \left(X_{i_{N-q-1}} A_{d,i_{N-q-1}}\right)^{\mathrm{T}} & \left(Z_{i_{N-q-1}} A_{d,i_{N-q-1}}\right)^{\mathrm{T}} \\ \hdashline 0 & 0 \end{array} \right],$$

$$\Xi_{\mathbb{Z}_{[1,N-1]},25} = \left[\begin{array}{cccccc} 0 & 0 & 0 & 0 & 0 & (H_{i_{N-q-1}}^d)^{\mathrm{T}} \\ 0 & 0 & 0 & 0 & 0 & 0 \end{array} \right],$$

$$\Xi_{\mathbb{Z}_{[1,N-1]},33} = \left[\begin{array}{c:c} -\gamma^2 \mathbf{I} & F_{i_{N-q-1}}^{\mathrm{T}} - D_{i_{N-q-1}}^{\mathrm{T}} D_{F,i_{N-q-1}}^{\mathrm{T}} \\ \hdashline * & -\mathbf{I} \end{array} \right],$$

$$\Xi_{\mathbb{Z}_{[1,N-1]},34} = \left[\begin{array}{c:c} \begin{array}{c} \left(X_{i_{N-q-1}} B_{i_{N-q-1}} \right. \\ \left. + W_{i_{N-q-1}} D_{i_{N-q-1}}\right)^{\mathrm{T}} \end{array} & \begin{array}{c} \left(Z_{i_{N-q-1}} B_{i_{N-q-1}} \right. \\ \left. + W_{i_{N-q-1}} D_{i_{N-q-1}}\right)^{\mathrm{T}} \end{array} \\ \hdashline 0 & 0 \end{array} \right],$$

$$\Xi_{\mathbb{Z}_{[1,N-1]},44} = \left[\begin{array}{c:c} \begin{array}{c} -X_{i_{N-q-1}} - X_{i_{N-q-1}}^{\mathrm{T}} \\ + \tilde{M}_{i_0, i_{N-q},1}^{f_{N-q-1}, f_{N-q}} \end{array} & \begin{array}{c} -Y_{i_{N-q-1}} - Z_{i_{N-q-1}}^{\mathrm{T}} \\ + \tilde{M}_{i_0, i_{N-q},2}^{f_{N-q-1}, f_{N-q}} \end{array} \\ \hdashline * & \begin{array}{c} -Y_{i_{N-q-1}} - Y_{i_{N-q-1}}^{\mathrm{T}} \\ + \tilde{M}_{i_0, i_{N-q},3}^{f_{N-q-1}, f_{N-q}} \end{array} \end{array} \right],$$

$$\Xi_{\mathbb{Z}_{[1,N-1]},45} = \left[\begin{array}{cccccc} X_{i_{N-q-1}} E_{i_{N-q-1}} & 0 & X_{i_{N-q-1}} E_{i_{N-q-1}} & 0 & 0 & 0 \\ Z_{i_{N-q-1}} E_{i_{N-q-1}} & 0 & Z_{i_{N-q-1}} E_{i_{N-q-1}} & 0 & 0 & 0 \end{array} \right],$$

$$\Xi_{\mathbb{Z}_{[1,N-1]},55} = \mathrm{diag}_6(-\varepsilon \mathbf{I}),$$

$$\Xi_{0,11} = \left[\begin{array}{c:c} -P_{i_0,1}^{f_0} + Q_1 & -P_{i_0,2}^{f_0} + Q_2 \\ \hdashline * & -P_{i_0,3}^{f_0} + Q_3 \end{array} \right],$$

$$\Xi_{0,13} = \left[\begin{array}{c:c} 0 & G_{i_0}^{\mathrm{T}} - C_{i_0}^{\mathrm{T}} D_{F,i_0}^{\mathrm{T}} \\ \hdashline 0 & -C_{F,i_0}^{\mathrm{T}} \end{array} \right],$$

$$\Xi_{0,14} = \left[\begin{array}{c:c} \begin{array}{c} \left(X_{i_0} A_{i_0}\right)^{\mathrm{T}} \\ + \left(W_{i_0} C_{i_0}\right)^{\mathrm{T}} \end{array} & \begin{array}{c} \left(Z_{i_0} A_{i_0}\right)^{\mathrm{T}} \\ + \left(W_{i_0} C_{i_0}\right)^{\mathrm{T}} \end{array} \\ \hdashline V_{i_0}^{\mathrm{T}} & V_{i_0}^{\mathrm{T}} \end{array} \right],$$

$$\Xi_{0,15} = \left[\begin{array}{cccccc} 0 & 0 & 0 & (H_{i_0}^a)^{\mathrm{T}} & 0 & 0 \\ 0 & 0 & 0 & 0 & 0 & 0 \end{array} \right],$$

$$\Xi_{0,22} = \left[\begin{array}{cc} -Q_1 & -Q_2 \\ * & -Q_3 \end{array} \right],$$

$$\Xi_{0,25} = \left[\begin{array}{cccccc} 0 & 0 & 0 & 0 & 0 & (H_{i_0}^d)^{\mathrm{T}} \\ 0 & 0 & 0 & 0 & 0 & 0 \end{array} \right],$$

$$\Xi_{0,33} = \left[\begin{array}{c|c} -\gamma^2 I & F_{i_0}^T - D_{i_0}^T D_{F,i_0}^T \\ \hline * & -I \end{array} \right],$$

$$\Xi_{N,34} = \left[\begin{array}{c|c} (X_{i_0} B_{i_0})^T & (Z_{i_0} B_{i_0})^T \\ + (W_{i_0} D_{i_0})^T & + (W_{i_0} D_{i_0})^T \\ \hline 0 & 0 \end{array} \right],$$

$$\Xi_{N,44} = \left[\begin{array}{c|c} -X_{i_0} - X_{i_0}^T & -Y_{i_0} - Z_{i_0}^T \\ + \tilde{M}_{i_0,i_1,1}^{f_0,f_1} & + \tilde{M}_{i_0,i_1,2}^{f_0,f_1} \\ \hline & -Y_{i_0} - Y_{i_0}^T \\ * & + \tilde{M}_{i_0,i_1,3}^{f_0,f_q} \end{array} \right],$$

$$\Xi_{N,45} = \left[\begin{array}{cccccc} X_{i_0} E_{i_0} & 0 & X_{i_0} E_{i_0} & 0 & 0 & 0 \\ Z_{i_0} E_{i_0} & 0 & Z_{i_0} E_{i_0} & 0 & 0 & 0 \end{array} \right],$$

$$\Xi_{N,55} = \mathrm{diag}_6(-\varepsilon I),$$

and $\tilde{M}_{i_0,i_{N-1}}^{f_{N-2},f_{N-1}}$, $\tilde{P}_{i_0,i_N}^{f_{N-1},f_N}$, $\tilde{M}_{i_0,i_{N-q-1}}^{f_{N-q-2},f_{N-q-1}}$, $\tilde{M}_{i_0,i_{N-q}}^{f_{N-q-1},f_{N-q}}$, $\tilde{M}_{i_0,i_1}^{f_0,f_1}$ are as in Lemma 10.1. Moreover, if (10.42a)–(10.42c) hold, then the desired filtering parameters of the form (10.5) can be obtained as

$$A_{F,i_0} = Y_{i_0}^{-1} V_{i_0}, B_{F,i_0} = Y_{i_0}^{-1} W_{i_0}, C_{F,i_0}, D_{F,i_0}. \tag{10.43}$$

Proof. Noting that $-R_{i_{N-1}} (\tilde{P}_{i_0,i_N}^{f_{N-1},f_N})^{-1} R_{i_{N-1}}^T \le -R_{i_{N-1}} - R_{i_{N-1}}^T + \tilde{P}_{i_0,i_N}^{f_{N-1},f_N}$, applying the well-known Schur complement [33] to (10.10a), and performing a congruence transformation $\mathrm{diag}\{I,I,I,I,R_{i_{N-1}}\}$, $\mathrm{diag}\{I,I,I,I,R_{i_{N-1}}^T\}$ to (10.10a), we have

$$\left[\begin{array}{ccccc} -\tilde{M}_{i_0,i_{N-1}}^{f_{N-2},f_{N-1}} + Q & 0 & 0 & \hat{C}_{i_{N-1}}^T & (R_{i_{N-1}} \hat{A}_{i_{N-1}})^T \\ * & -Q & 0 & 0 & (R_{i_{N-1}} \hat{A}_{d,i_{N-1}})^T \\ * & * & -\gamma^2 I & \hat{D}_{i_{N-1}}^T & (R_{i_{N-1}} \hat{B}_{i_{N-1}})^T \\ * & * & * & -I & 0 \\ * & * & * & * & -R_{i_{N-1}} - R_{i_{N-1}}^T + \tilde{P}_{i_0,i_N}^{f_{N-1},f_N} \end{array} \right] < 0. \tag{10.44}$$

Letting

$$V_{i_{N-1}} = Y_{i_{N-1}} A_{F,i_{N-1}}, W_{i_{N-1}} = Y_{i_{N-1}} B_{F,i_{N-1}},$$

$$\tilde{M}_{i_0,i_{N-1}}^{f_{N-2},f_{N-1}} = \left[\begin{array}{cc} \tilde{M}_{i_0,i_{N-1},1}^{f_{N-2},f_{N-1}} & \tilde{M}_{i_0,i_{N-1},2}^{f_{N-2},f_{N-1}} \\ * & \tilde{M}_{i_0,i_{N-1},3}^{f_{N-2},f_{N-1}} \end{array} \right],$$

$$\tilde{P}_{i_0,i_N}^{f_{N-1},f_N} = \begin{bmatrix} \tilde{P}_{i_0,i_N,1}^{f_{N-1},f_N} & \tilde{P}_{i_0,i_N,2}^{f_{N-1},f_N} \\ * & \tilde{P}_{i_0,i_N,3}^{f_{N-1},f_N} \end{bmatrix},$$

$$R_{i_{N-1}} = \begin{bmatrix} X_{i_{N-1}} & Y_{i_{N-1}} \\ Z_{i_{N-1}} & Y_{i_{N-1}} \end{bmatrix}, Q = \begin{bmatrix} Q_1 & Q_2 \\ * & Q_3, \end{bmatrix} \qquad (10.45)$$

and substituting (10.45) into (10.44), we have

$$\Upsilon_N = \begin{bmatrix} \Xi_{N,11} & 0 & \Xi_{N,13} & \Upsilon_{N,14} \\ * & \Xi_{N,22} & 0 & \Upsilon_{N,24} \\ * & * & \Xi_{N,33} & \Xi_{N,34} \\ * & * & * & \Xi_{N,44} \end{bmatrix} < 0, \qquad (10.46)$$

where

$$\Upsilon_{N,14} = \begin{bmatrix} \left(X_{i_{N-1}}\tilde{A}_{i_{N-1}}\right)^{\mathrm{T}} \\ + \left(W_{i_{N-1}}C_{i_{N-1}}\right)^{\mathrm{T}} & \left(Z_{i_{N-1}}\tilde{A}_{i_{N-1}}\right)^{\mathrm{T}} \\ + \left(W_{i_{N-1}}C_{i_{N-1}}\right)^{\mathrm{T}} \\ V_{i_{N-1}}^{\mathrm{T}} & V_{i_{N-1}}^{\mathrm{T}} \end{bmatrix},$$

$$\Upsilon_{N,24} = \begin{bmatrix} \left(X_{i_{N-1}}\tilde{A}_{d,i_{N-1}}\right)^{\mathrm{T}} & \left(Z_{i_{N-1}}\tilde{A}_{d,i_{N-1}}\right)^{\mathrm{T}} \\ 0 & 0 \end{bmatrix}.$$

By taking (10.2) into account this further yields

$$\Upsilon_N = \Psi_N + \begin{bmatrix} 0 & 0 & 0 \\ \vdots & \vdots & \vdots \\ X_{i_{N-1}}E_{i_{N-1}} & 0 & X_{i_{N-1}}E_{i_{N-1}} \\ X_{i_{N-1}}E_{i_{N-1}} & 0 & X_{i_{N-1}}E_{i_{N-1}} \end{bmatrix}$$

$$\times \mathscr{L}_{i_{N-1}}(k) \begin{bmatrix} H_{i_{N-1}}^a & 0 & 0 & 0 & \cdots & 0 \\ 0 & 0 & 0 & 0 & \cdots & 0 \\ 0 & 0 & H_{i_{N-1}}^d & 0 & \cdots & 0 \end{bmatrix}, \qquad (10.47)$$

where

$$\Psi_N = \begin{bmatrix} \Xi_{N,11} & 0 & \Xi_{N,13} & \Psi_{N,14} \\ * & \Xi_{N,22} & 0 & \Psi_{N,24} \\ * & * & \Xi_{N,33} & \Xi_{N,34} \\ * & * & * & \Xi_{N,44} \end{bmatrix} < 0,$$

$$\Psi_{N,14} = \begin{bmatrix} \left(X_{i_{N-1}} A_{i_{N-1}}\right)^{\mathrm{T}} & \vdots & \left(Z_{i_{N-1}} A_{i_{N-1}}\right)^{\mathrm{T}} \\ + \left(W_{i_{N-1}} C_{i_{N-1}}\right)^{\mathrm{T}} & \vdots & + \left(W_{i_{N-1}} C_{i_{N-1}}\right)^{\mathrm{T}} \\ \hline V_{i_{N-1}}^{\mathrm{T}} & \vdots & V_{i_{N-1}}^{\mathrm{T}} \end{bmatrix},$$

$$\Psi_{N,24} = \begin{bmatrix} \left(X_{i_{N-1}} A_{d,i_{N-1}}\right)^{\mathrm{T}} & \vdots & \left(Z_{i_{N-1}} A_{d,i_{N-1}}\right)^{\mathrm{T}} \\ \hline 0 & \vdots & 0 \end{bmatrix}.$$

By applying the Schur complement and Lemma 10.3 to (10.47), we have that $\Upsilon_N < 0$ implies $\Xi_N < 0$. Therefore, we can conclude that condition (10.10a) can be transformed into an LMI formulation (10.42a) with designed filter gains. Following the lines of the proof of (10.42a), we can prove that the filtering gain matrices obtained by Theorem 10.1 can achieve a bounded filtering error (10.9). This completes the proof. □

10.5 NUMERICAL EXAMPLE

In this section, we provide an illustrative example to verify the effectiveness and applicability of the proposed NALKF method.

Consider NMJDS (10.1) with the following parameters:

$$A_1 = \begin{bmatrix} 0.5 & -0.6 \\ -0.4 & -0.2 \end{bmatrix}, A_{d,1} = \begin{bmatrix} -0.2 & 0.1 \\ 0.2 & 0.15 \end{bmatrix}, B_1 = \begin{bmatrix} 0.4 \\ 0.5 \end{bmatrix},$$

$$A_2 = \begin{bmatrix} 0.1 & 0.2 \\ 0.15 & 0.3 \end{bmatrix}, A_{d,2} = \begin{bmatrix} 0.38 & 0 \\ 0.01 & 0.16 \end{bmatrix}, B_2 = \begin{bmatrix} 0.2 \\ 0.6 \end{bmatrix},$$

$$C_1 = \begin{bmatrix} 0.1 \\ -0.1 \end{bmatrix}, D_1 = 0.4, G_1 = \begin{bmatrix} 0.7 & 0.3 \end{bmatrix}, F_1 = 0.2,$$

$$C_2 = \begin{bmatrix} 0.3 \\ -0.4 \end{bmatrix}, D_2 = -0.5, G_2 = \begin{bmatrix} 0.2 & 0.4 \end{bmatrix}, F_2 = -0.1,$$

$$E = \begin{bmatrix} 0.2 & 0 \\ 0 & 0.1 \end{bmatrix}, H_1^a = \begin{bmatrix} 0.03 & 0 \\ 0 & 0.04 \end{bmatrix}, H_2^a = \begin{bmatrix} 0.02 & 0 \\ 1 & 0.05 \end{bmatrix},$$

$$H_1^d = \begin{bmatrix} 0.01 & 0 \\ 0 & 0.02 \end{bmatrix}, H_2^d = \begin{bmatrix} 0.32 & 0 \\ 1 & 0.68 \end{bmatrix},$$

$$\mathscr{L}_1(k) = \mathscr{L}_2(k) = \begin{bmatrix} \sin(k) & 0 \\ 0 & \sin(k) \end{bmatrix}.$$

The TP matrix is assumed to be time-varying in a polytope given by its vertices

$$\Pi_1 = \begin{bmatrix} 0.1 & 0.9 \\ 0.2 & 0.8 \end{bmatrix}, \Pi_2 = \begin{bmatrix} 0.3 & 0.7 \\ 0.4 & 0.6 \end{bmatrix}.$$

The indicator functions are chosen as

$$\mu_1(k) = \frac{1 - x_{k,1}}{2}, \mu_2(k) = \frac{1 + x_{k,1}}{2}.$$

where $x_{k,1}$ represents the first state of x_k. The initial state, initial estimated state, initial z_k and initial $\hat{z}_k$ are taken as $x_0 = \hat{x}_0 = [0 \quad 0]^T$, $z_0 = \hat{z}_0 = [0 \quad 0]^T$. The initial operating mode and time-delay are assumed as $r_0 = 1$ and $d = 4$. The exogenous disturbance w_k are taken as $0.2e^{-0.5k}$.

In this case, we would like to provide an optimal H_∞ performance for designing the filter, that is, we are interested in the optimization problem

$$\min \gamma \quad \text{subject to} \quad (10.42a)\text{--}(10.42c)$$

under the zero initial conditions. Solving (10.42a)–(10.42c) by using the LMI ToolBox, the robust H_∞ filter gains obtained via the two-step ahead approach are listed as

$$A_{F,1} = \begin{bmatrix} -0.0214 & 0.2915 \\ -0.0636 & -0.6562 \end{bmatrix}, A_{F,2} = \begin{bmatrix} 1.2150 & 0.3716 \\ -0.1274 & 0.1425 \end{bmatrix},$$

$$B_{F,1} = \begin{bmatrix} -3.2391 \\ -0.9403 \end{bmatrix}, B_{F,2} = \begin{bmatrix} 0.1913 \\ 1.5089 \end{bmatrix},$$

$$C_{F,1} = \begin{bmatrix} -0.6039 & -0.4670 \end{bmatrix}, C_{F,2} = \begin{bmatrix} -0.1541 & -0.4506 \end{bmatrix},$$

$$D_{F,1} = -0.5960, D_{F,2} = -0.1448.$$

Similarly, the robust H_∞ filter gains obtained via the three-step ahead approach are listed as

$$A_{F,1} = \begin{bmatrix} -0.3931 & 0.3412 \\ 0.7966 & -0.9720 \end{bmatrix}, A_{F,2} = \begin{bmatrix} 0.4723 & 0.3298 \\ 0.4882 & 0.3683 \end{bmatrix},$$

$$B_{F,1} = \begin{bmatrix} -7.3667 \\ 6.3261 \end{bmatrix}, B_{F,2} = \begin{bmatrix} 0.6209 \\ 0.6633 \end{bmatrix},$$

$$C_{F,1} = \begin{bmatrix} -0.5337 & -0.3707 \end{bmatrix}, C_{F,2} = \begin{bmatrix} -0.3412 & -0.0273 \end{bmatrix},$$

$$D_{F,1} = -0.9465, D_{F,2} = 0.3611.$$

TABLE 10.1   Minimum value of γ and number of the decision variables.

	Minimum value of γ	Number of decision variables
Conventional approach	8.7815	81
Two-step approach	1.8374	117
Three-step approach	1.6383	117

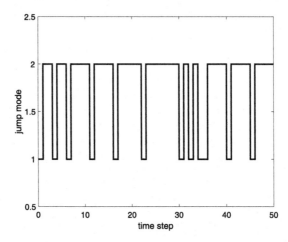

FIGURE 10.1   Jumping mode path.

We must point out that the conventional filtering approach, such as in [18], can be viewed as a particular case of the NALKF approach, namely, the one-step ahead approach. The filtering gains can be obtained as

$$A_{F,1} = \begin{bmatrix} 0.0531 & -0.0076 \\ -0.0727 & -0.5311 \end{bmatrix}, A_{F,2} = \begin{bmatrix} 0.1398 & 1.7352 \\ 0.0533 & 0.7602 \end{bmatrix},$$

$$B_{F,1} = \begin{bmatrix} -9.8681 \\ 4.2420 \end{bmatrix}, B_{F,2} = \begin{bmatrix} -1.3763 \\ -0.5301 \end{bmatrix},$$

$$C_{F,1} = \begin{bmatrix} 0.0761 & -1.0295 \end{bmatrix}, C_{F,2} = \begin{bmatrix} -0.0433 & -0.6545 \end{bmatrix},$$

$$D_{F,1} = -6.9897, D_{F,2} = -0.5292.$$

We also get the minimum value of γ and the number of decision variables of NALKF approach, which are shown in Table 10.1. Obviously, with the increase of the predictive step, we obtain better H_∞ performance. However, we must point out that the number of decision variables of the developed NALKF approach do not increase with the predictive step because the knowledge of the multistep TPs $\pi_{i_0,i_{N-q}}$ is included.

Jumping mode paths shown in Fig. 10.1 from time step 0 to time step 50 are generated randomly by employing TP matrices Π_1 and Π_2 and indi-

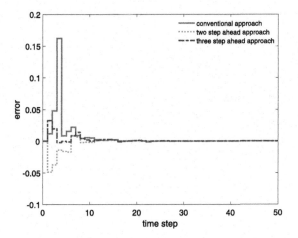

FIGURE 10.2 Estimated error.

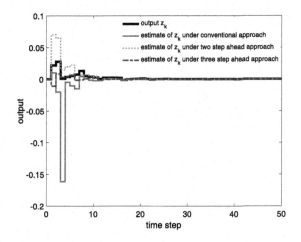

FIGURE 10.3 Estimate of z_k.

cator functions $\mu_1(k)$ and $\mu_2(k)$. The number of iterations is chosen as 50, and each iteration unit length is taken as 1.

The comparison of the estimated error is demonstrated in Fig. 10.2, where the red (mid gray in print version), green (light gray in print version), and blue (dark gray in print version) lines represent the error under conventional filtering approach, two-step ahead approach, and three-step ahead approach, respectively. We can also conclude that with the increase of the predictive step, the estimated error becomes smaller.

Moreover, the estimate of the output $\hat{z}_k$ is also drawn in Fig. 10.3, where the black line is the output z_k. We can observe that the three-step ahead

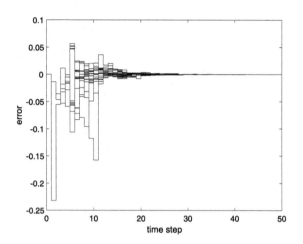

FIGURE 10.4 Monte Carlo simulation of estimated error.

approach (blue line; dark gray in print version) has the best tracking performance, which also justifies the results of Fig. 10.2.

Finally, a Monte Carlo simulation of the estimated error, which is generated under 20 different jumping mode paths, is given in Fig. 10.4. Such a simulation uses the two-step ahead approach. We can see that the overall estimation quality is good. The extreme case is less than 25%. Therefore, it justifies the effectiveness of the developed approach from another side.

10.6 CONCLUSION

The robust H_∞ filtering problem has been considered in this chapter for nonhomogeneous Markovian jump delay systems via the N-step ahead Lyapunov–Krasovskii function approach. The robust H_∞ filter has been designed in terms of a feasible optimization problem subject to LMI constraints, which guarantees the filtering error system to be mean-square stable and the filtering error to satisfy a smaller H_∞ dissipative level for all possible time-varying transition probabilities and all admissible parameter uncertainties and time delays. The N-step ahead approach can be extended to deal with the robust H_∞ dynamic output feedback control problem.

References

[1] X.H. Chang, Robust Output Feedback H_∞ Control and Filtering for Uncertain Linear Systems, Springer, London, 2014.

[2] R.E. Kalman, A new approach to linear filtering and prediction problems, Journal of Basic Engineering 82 (1) (1960) 35–45.

[3] M.S. Mahmoud, N.F. Al-Muthairi, S. Bingulac, Robust Kalman filtering for continuous time-lag systems, Systems & Control Letters 38 (4–5) (1999) 309–319.

[4] X. Zhu, Y.C. Soh, L. Xie, Design and analysis of discrete-time robust Kalman filters, Automatica 38 (6) (2002) 1069–1077.

[5] M.S. Mahmoud, H.M. Khalid, Distributed Kalman filtering: a bibliographic review, IET Control Theory & Applications 7 (4) (2013) 483–501.

[6] W. Li, Y. Jia, J. Du, Distributed extended Kalman filter with nonlinear consensus estimate, Journal of the Franklin Institute 354 (17) (2017) 7983–7995.

[7] G. Battistelli, L. Chisci, D. Selvi, A distributed Kalman filter with event-triggered communication and guaranteed stability, Automatica 93 (2018) 75–82.

[8] D. Viegas, P. Batista, P. Oliveira, et al., Discrete-time distributed Kalman filter design for formations of autonomous vehicles, Control Engineering Practice 75 (2018) 55–68.

[9] S. Dey, A.S. Leong, J.S. Evans, Kalman filtering with faded measurements, Automatica 45 (10) (2009) 2223–2233.

[10] H. Geng, Z. Wang, Y. Liang, et al., Tobit Kalman filter with fading measurements, Signal Processing 140 (2017) 60–68.

[11] X. Liu, L. Li, Z. Li, et al., Stochastic stability of modified extended Kalman filter over fading channels with transmission failure and signal fluctuation, Signal Processing 138 (2017) 220–232.

[12] H. Gao, X. Li, Robust Filtering for Uncertain Systems-A Parameter-Dependent Approach, Springer, London, 2014.

[13] Y. Liu, G. Yang, X. Li, Event-based robust H_∞ filtering for affine fuzzy systems, Fuzzy Sets and Systems 329 (2017) 19–35.

[14] X.Q. Xiao, L. Zhou, G.P. Liu, Event-triggered H_∞ filtering of continuous-time switched linear systems, Signal Processing 141 (2017) 343–349.

[15] X. Xiao, J.H. Park, L. Zhou, Event-triggered H_∞ filtering of discrete-time switched linear systems, ISA Transactions 77 (2018) 112–121.

[16] J. Wang, F. Li, Y. Sun, Asynchronous $l_2 - l_\infty$ filtering for networked fuzzy systems with Markov jump parameters over a finite-time interval, IET Control Theory & Applications 10 (17) (2016) 2175–2185.

[17] L.X. Zhang, T. Yang, P. Shi, Analysis and Design of Markov Jump Systems with Complex Transition Probabilities, Springer, Switzerland, 2016.

[18] H. Liu, Y. Ding, J. Cheng, New results on H_∞ filtering for Markov jump systems with uncertain transition rates, ISA Transactions 69 (2017) 43–50.

[19] M. Sathishkumar, R. Sakthivel, C. Wang, et al., Non-fragile filtering for singular Markovian jump systems with missing measurements, Signal Processing 142 (2018) 125–136.

[20] Y.Y. Yin, P. Shi, F. Liu, et al., Filtering for discrete-time nonhomogeneous Markov jump systems with uncertainties, Information Sciences 259 (2014) 118–127.

[21] M. Hua, L. Zhang, F. Yao, et al., Robust H_∞ filtering for continuous-time nonhomogeneous Markov jump nonlinear systems with randomly occurring uncertainties, Signal Processing 148 (2018) 250–259.

[22] J. Wang, S. Ma, C. Zhang, Finite-time H_∞ filtering for nonlinear singular systems with nonhomogeneous Markov jumps, IEEE Transactions on Cybernetics (2018), Early Access, https://doi.org/10.1109/TCYB.2018.2820139.

[23] A.A. Ahmadi, P.A. Parrilo, Non-monotonic Lyapunov functions for stability of discrete time nonlinear and switched systems, in: Proceedings of 47th IEEE Conference on Decision and Control, Cancun, Mexico, 2008, pp. 614–621.

[24] S.F. Derakhshan, A. Fatehi, Non-monotonic Lyapunov functions for stability analysis and stabilization of discrete time Takagi–Sugeno fuzzy systems, International Journal of Innovative Computing, Information & Control (2014) 1567–1586.

[25] S.F. Derakhshan, A. Fatehi, M.G. Sharabiany, Nonmonotonic observer-based fuzzy controller designs for discrete time T-S fuzzy systems via LMI, IEEE Transactions on Cybernetics 44 (12) (2014) 2557–2567.

[26] S.F. Derakhshan, A. Fatehi, Non-monotonic robust H_2 fuzzy observer-based control for discrete time nonlinear systems with parametric uncertainties, International Journal of Systems Science 4 (12) (2015) 2134–2149.

[27] A. Kruszewski, T.M. Guerra, S. Labiod, Stabilization of Takagi–Sugeno discrete models: towards an unification of the results, in: IEEE International Fuzzy Systems Conference, 2007, pp. 1–6.

[28] A. Kruszewski, R. Wang, T.M. Guerra, Non-quadratic stabilization conditions for a class of uncertain nonlinear discrete time T-S fuzzy models: a new approach, IEEE Transactions on Automatic Control 53 (2) (2008) 606–611.

[29] T.M. Guerra, A. Kruszewski, S. Lauber, Discrete Tagaki–Sugeno models for control: where are we?, Annual Reviews in Control 33 (1) (2009) 37–47.

[30] T.M. Guerra, A. Kruszewski, M. Bernal, Control law proposition for the stabilization of discrete Takagi–Sugeno models, IEEE Transactions on Fuzzy Systems 17 (3) (2009) 724–731.

[31] Y.J. Chen, M. Tanaka, K. Inoue, H. Ohtake, K. Tanaka, T.M. Guerra, A. Kruszewski, H.O. Wang, A non-monotonically decreasing relaxation approach of Lyapunov functions to guaranteed cost control for discrete fuzzy systems, IET Control Theory & Applications 8 (16) (2014) 1716–1722.

[32] A. Nasiri, S.K. Nguang, A. Swain, D.J. Almakhles, Robust output feedback controller design of discrete-time Takagi–Sugeno fuzzy systems: a non-monotonic Lyapunov approach, IET Control Theory & Applications 10 (5) (2016) 545–553.

[33] A. Nasiri, S.K. Nguang, A. Swain, Reducing conservatism in an H_∞ robust state-feedback control design of T-S fuzzy systems: a non-monotonic approach, IEEE Transactions on Fuzzy Systems 26 (1) (2018) 386–390.

[34] J. Wen, S.K. Nguang, P. Shi, X.D. Zhao, Stability and H_∞ control of discrete-time switched systems via one-step ahead Lyapunov function approach, IET Control Theory & Applications 12 (8) (2018) 1141–1147.

[35] Y. Xie, J. Wen, S.K. Nguang, et al., Stability, l_2-gain and robust H_∞ control for switched systems via N-step ahead Lyapunov function approach, IEEE Access 5 (1) (2017) 26400–26408.

[36] J. Wen, S.K. Nguang, P. Shi, et al., Robust H_∞ control of discrete-time non-homogenous Markovian jump systems via multistep Lyapunov function approach, IEEE Transactions on Systems, Man, and Cybernetics: Systems 47 (7) (2017) 1439–1450.

[37] S. Boyd, L.E. Ghaoui, E. Feron, V. Balakrishnan, Linear Matrix Inequalities in System and Control Theory, SIAM Studies in Applied Mathematics, 1994.

[38] S. Zhou, G. Feng, H_∞ filtering for discrete-time systems with randomly varying sensor delays, Automatica 44 (7) (2008) 1918–1922.

[39] Y. Zhang, G. Cheng, C. Liu, Finite-time unbiased H_∞ filtering for discrete jump time-delay systems, Applied Mathematical Modelling 38 (13) (2014) 3339–3349.

[40] F. Wang, W. Che, H. Xu, H_∞ filtering for uncertain systems with time-delay and randomly occurred sensor nonlinearities, Neurocomputing 174 (2016) 571–576.

[41] Y. Cao, Z. Lin, T. Hu, Stability analysis of linear time-delay systems subject to input saturation, IEEE Transactions on Circuits and Systems. I, Fundamental Theory and Applications 49 (2) (2002) 233–240.

11

Conclusions and Future Work

This chapter draws conclusions on the book and points out some possible research directions related to the work done in this book.

11.1 CONCLUSIONS

The focus of the book has been placed on robust control and filtering for T-S fuzzy model, switched systems, and nonhomogeneous Markovian jump systems. Several research problems have been investigated in detail.

First, we aim at T-S fuzzy models with modeling uncertainties. By properly constructing LF with nonmonotonic behavior, the criteria for stability analysis, stabilization, and filtering are presented and then are extended to the case of dynamic output feedback control of nonlinear systems.

A nonmonotonic Lyapunov function (NLF) has been employed to design robust H_∞ state feedback controllers for uncertain T-S fuzzy systems in Chapter 2. In the NLF approach, the monotonicity requirement of the LF is relaxed by allowing it to increase locally but go to zero ultimately. Based on the NLF approach, sufficient conditions for the existence of a robust state feedback H_∞ controller that guarantee the stability and a prescribed H_∞ performance have been provided in terms of LMIs. In Chapter 3, we have examined the robust H_∞ filtering problem for T-S fuzzy models of nonlinear systems by using the idea of nonmonotonic approach. Our main contribution has further reduced the conservatism and improved the H_∞ performance. Motivated by the results in Chapter 2, in Chapter 4, robust H_∞ output feedback control for uncertain T-S fuzzy systems has been studied via NLF approaches. For comparison, the proposed design technique has been shown to be less conservative than the existing nonmonotonic approach, namely, N-sample variations of LFs.

Second, we focus on arbitrary switched systems and average dwell-time switched systems. Methodologies that can effectively handle control and filtering problems with less conservatism are developed by allowing the LF to increase both at the switching instant and during the running time of each subsystem.

Non-Monotonic Approach to Robust H∞ Control of Multi-Model Systems
https://doi.org/10.1016/B978-0-12-814868-6.00017-4

Chapter 5 has been focused on the H_∞ control for a class of discrete-time switched systems by introducing the one-step ahead LF approach. The one-step ahead LF is a function of future states. The design objectives are to reduce the conservatism of the stability criterion developed for arbitrarily switched systems and further get a better disturbance attenuation capability. The distinguishing feature is that the one-step ahead LF has no structural constraint, such as a diagonal structure, and the resulting analysis and synthesis criteria can cover the nonmonotonic method considering two-sample variation as a particular case. In Chapter 6, we have developed an N-step ahead LF approach, which allows a nonmonotonic behavior both at the switching instants and during the running time of each subsystem. The asymptotic stability criterion has been improved as well as the capability of disturbance attenuation. By introducing a series of auxiliary variables the future knowledge of states and exogenous noises can be properly used to derive sufficient conditions for the existence of a robust H_∞ controller in the form of a set of numerical testable conditions. Moreover, the relationship between the N-step time difference of LF and switching rate, i.e., the ADT constraint, is thoroughly discussed. In Chapter 7, the nonmonotonic Lyapunov function approach with N-step ahead predictive horizon has been developed for designing a robust H_∞ filter for a discrete-time uncertain switched system. With increasing the number N, the filtering performance can be improved as well as the capability of disturbance attenuation. To further relax the restrictions on the switching law, the mode-dependent ADT switching has been introduced to reduce the ADT bound such that a trade-off between the switching frequency and filtering performance can be achieved. Chapter 8 developed dissipative dynamic output feedback (DOF) control for a class of average dwell-time switching systems via the multistep LF approach. First, a larger dissipative region with guaranteed stability and, specifically, smaller H_∞ level can be achieved by increasing a predictive step N, which means that the monotonic requirement of LF is relaxed. Then, based on the results of dissipative analysis, a robust dissipative DOF controller is further designed.

Finally, we deal with Markovian jump systems governed by time-varying transition probabilities. The concept of N-step ahead approach is proposed such that the stability problem can be solved via a finite number of conditions. The systems involved with time-delay, which are handled by the Lyapunov–Krasovskii function approach, are also investigated.

In Chapter 9, the robust H_∞ control problem for a class of discrete-time nonhomogenous Markovian jump linear systems (NMJLSs) has been investigated by a multistep LF approach. The proposed multistep LF is allowed to increase during the period of several sampling time steps ahead of the current time within the jump mode. First, a less conservative stability criterion has been derived based on this multistep LF approach. Second, an H_∞ performance has been analyzed under the multistep case by properly dealing with the knowledge of future states and extraneous noises. These two results have then been employed to facilitate a robust H_∞ control design for NMJLSs. Chapter 10 studied the robust H_∞ filtering problem for a class of nonhomogeneous Markovian jump delay systems with the N-step ahead Lyapunov–Krasovskii function approach. In this chapter, we aim to design filters such that, for all possible time-varying transition probabilities and all admissible parameter uncertainties and time-delays, the filtering error system is mean-square stable with a smaller estimated error and a lower dissipative level. In terms of LMIs, sufficient conditions for the solvability of the addressed problem are developed via a moving-horizon method to avoid essential difficulties introduced by future noise.

11.2 FUTURE WORK

Some of future works are listed as follows:

- A possible future research direction is to investigate multiobjective H_2–H_∞ control and filtering problems for multimodel systems.
- It would be interesting to investigate the problems of fault detection and fault tolerant control for multimodel systems via the nonmonotonic Lyapunov function approach.
- A trend for future research is to generalize the methods obtained in the book to the finite-time or finite-frequency control and filtering problems.
- A challenging work is to employ the nonmonotonic approach to the model predictive control. A possible difficulty is to make a trade-off between the conservatism reduction and online computational burden.
- The nonlinearities considered in the book have some condition constraints that bring conservativeness. The analysis and synthesis of general nonlinear systems could be one of the future research works.

Index

Printed in the United States
By Bookmasters